Branching Processes

ADVANCES IN PROBABILITY and Related Topics

Editor: Peter Ney

Department of Mathematics
University of Wisconsin-Madison
Madison, Wisconsin

Vols. 1 and 2 Advances in Probability and Related Topics, edited by Peter Ney

Vol. 3 Advances in Probability and Related Topics, edited by Peter Ney and Sidney Port

Vol. 4 Probability on Banach Spaces, edited by James Kuelbs

Vol. 5 Branching Processes, edited by Anatole Joffe and Peter Ney

Vol. 6 Multicomponent Random Systems, edited by R. L. Dobrushin and Ya. G. Sinai (in preparation)

Other volumes in preparation

Branching Processes

edited by Anatole Joffe
Centre de recherches mathématiques
Université de Montréal
Montreal, Quebec, Canada

Peter Ney
Department of Mathematics
University of Wisconsin—Madison
Madison, Wisconsin

MARCEL DEKKER, INC. New York and Basel

Library of Congress Cataloging in Publication Data

Main entry under title:

Branching processes.

(Advances in probability and related topics ; v. 5)

Papers presented at a conference held in Quebec, August 11-20, 1976 under the sponsorship of the Centre de recherches mathématiques of the Université de Montréal and the National Research Council of Canada.

Includes indexes.

1. Branching processes--Congresses. I. Joffe, Anatole [Date]. II. Ney, Peter [Date]. III. Université de Montréal. Centre de recherches mathématiques. IV. National Research Council of Canada. V. Series.

QA273.A1A4 vol. 5 [QA274.76] 519.2'08s
ISBN 0-8247-6800-0 [519.2'34] 78-11727

Marcel Dekker, Inc.
270 Madison Avenue, New York, New York 10016

Current printing (last digit):
10 9 8 7 6 5 4 3 2 1

PRINTED IN THE UNITED STATES OF AMERICA

ABOUT THE SERIES

The *Advances in Probability* was founded in 1970 to provide a flexible medium for publication of work in probability theory and related areas. The idea was to encourage lucid exposition of topics of current interest, while giving authors greater flexibility on the inclusion of original and expository material and on the lengths of articles than might be available in standard journals. There has not been any particular focus of subject matter, other than its relevance to probability theory.

During the past decade, a variety of series have evolved which now offer the opportunity for this kind of publication. We have therefore decided to modify our format to focus the attention of each volume on a single unified subarea of probability theory, while retaining flexibility in the format of individual articles.

To this end, Volume 4 focused on "Probability on Banach Spaces," the present volume centers on "Branching Processes," while a planned volume will deal with "Multicomponent Random Systems."

We intend to maintain flexible editorial arrangements. The editors for the various volumes will in general be experts in a special area of probability theory and will be responsible for the contents of their particular volume. People who might be interested in editing such a volume are invited to contact

Peter Ney, Series Editor
Department of Mathematics
University of Wisconsin - Madison

PREFACE

On August 11 to 20, 1976, a small conference on branching processes was held in Quebec under the sponsorship of the Centre de recherches mathématiques of the Université de Montréal, and the National Research Council of Canada. In conjunction with the meeting, participants were invited to contribute papers to the present volume.

Obviously it was not possible to bring together all the active workers in the field, nor to represent all areas of activity. We note, for example, that there have been interesting developments on branching random walks and diffusions during the two years since the conference.

Nevertheless, many of the major lines of work of the past decade were discussed at the meeting, and are partially represented in this volume.

The participants in the conference were I. Ahmat, S. Asmussen, K. Athreya, M. Bramson, J. Chover, J. Corona, D. Dawson, J. P. Dion, S. Dubuc, S. Durham, W. Esty, J. Foster, M. Goldstein, L. Gorostiza, H. Hering, F. Hoppe, G. Ivamoff, A. Joffe, N. Kaplan, T. Kurtz, A. Moncayo, P. Ney, D. Sankoff, and E. Seneta.

We would like to express our thanks to M. Goldstein and S. Dubuc for their participation in the organization of the conference. Our particular thanks also go to Claire O'Reilly of the Centre, who was principally responsible for the detailed arrangements, and whose efforts were essential to the success of the conference.

Anatole Joffe
Peter Ney
April 1978

CONTENTS

CONTRIBUTORS

Søren Asmussen, Institute of Mathematical Statistics, University of Copenhagen, Copenhagen, Denmark

K. B. Athreya, Department of Applied Mathematics, Indian Institute of Science, Bangalore, India

D. Dawson, Carleton University, Ottawa, Ontario, Canada

Jean-Pierre Dion, Département de mathématiques, Université de Québec à Montréal, Montreal, Quebec, Canada

S. Dubuc, Département de mathématiques et statistiques, Université de Montréal, Montreal, Quebec, Canada

Luis G. Gorostiza, Centro de Investigación del Instituto Politécnico Nacional, Mexico D.F., Mexico

Heinrich Hering, Universität Regensburg and Københavns Universität, Copenhagen, Denmark

F. M. Hoppe,* Cornell University, Ithaca, New York

G. Ivanoff, Carleton University, Ottawa, Ontario, Canada

A. Joffe, Centre de recherches mathématiques, Université de Montréal, Montreal, Quebec, Canada

N. Kaplan, Biometry Branch, National Institute of Environmental Health Sciences, Research Triangle Park, North Carolina

Niels Keiding, Institute of Mathematical Statistics, University of Copenhagen, Copenhagen, Denmark

Thomas G. Kurtz, Department of Mathematics, University of Wisconsin-Madison, Madison, Wisconsin

*Permanent address: University of Alberta, Edmonton, Alberta, Canada

Alberto Ruíz Moncayo, Universidad Autónoma Metropolitana Iztapalapa Mexico D.F., Mexico

Peter Ney, Department of Mathematics, University of Wisconsin-Madison, Madison, Wisconsin

E. Seneta,* Virginia Polytechnic Institute and State University, Blacksburg, Virginia

---

*Permanent address: Australian National University, Canberra, Australia

Branching Processes

SOME MARTINGALE METHODS IN THE LIMIT THEORY OF SUPERCRITICAL BRANCHING PROCESSES

Søren Asmussen

Institute of Mathematical Statistics
University of Copenhagen
Copenhagen, Denmark

1. INTRODUCTION

Consider a Galton-Watson process $\{Z_n\}$ with offspring distribution F and offspring mean $m = \int_0^\infty x\,dF(x)$. We think of the process as constructed from a double sequence $\{X_{n,i}\}$ of independent random variables (r.v.) distributed according to F, such that

$$Z_0 = 1, \quad Z_{n+1} = \sum_{i=1}^{Z_n} X_{n,i} \ .$$

When $m \leq 1$, the only interesting a.s. statement on $\{Z_n\}$ seems to be the certainty of extinction and all limit results as well as their proofs are essentially analytic in nature. The flavour of the supercritical case $1 < m < \infty$, which we consider throughout, is quite different. Here in general growth to infinity occurs with positive probability, and the limit results are of strong type, describing for example the growth more precisely and more specific phenomena such as spatial distribution and age structure in the various generalizations of the model. Despite this fact, extensive use is made of analytic techniques, often succesfully and in a natural way, but often also in situations where the problems would suggest a different approach.

In the present paper, we present some probabilistic methods useful when dealing with certain aspects of the limit theory in the supercritical case. As is well-known, $W_n = Z_n/m^n$ is a non-negative martingale w.r.t. $\mathcal{F}_n = \sigma(X_{m,i};\ i = 1,2,3,\dots;\ m < n)$ and thus $W = \lim_n W_n$ exists [1]. Roughly speaking, our approach is to undertake a more refined study of $\{W_n\}$ and W in terms of the infinite series

$$(1.1)\qquad \sum_{n=0}^{\infty} \{W_{n+1} - W_n\} = \sum_{n=0}^{\infty} m^{-n-1} \sum_{i=1}^{Z_n} \{X_{n,i} - m\}$$

and to exploit the structure of (1.1) as something in between a general martingale series and a sum of independent r.v. with mean zero. The core of the paper is § 2-3, where we introduce some new martingale series approximating (1.1). The applications are in § 2 to give a new and short proof of the classical result of [13],

<u>1.1 Theorem</u> $EW = EW_0 = 1$ <u>if and only if</u>

$$(x \log x)\qquad \int_0^{\infty} x \log x \, dF(x) < \infty$$

<u>while</u> $W = 0$ <u>otherwise</u>.

and in § 3 to study convergence rates, i.e. a.s. estimates of $W - W_n$,

<u>1.2 Theorem</u> <u>Suppose</u> $(x \log x)$ <u>holds</u>. <u>Then</u>:

(i) Let $1 < p < 2$, $1/p + 1/q = 1$. <u>Then</u> $W - W_n = o(m^{-n/q})$ <u>if and only if</u> $\int_0^{\infty} x^p \, dF(x) < \infty$.

(ii) <u>Let</u> $\alpha > 0$. <u>Then</u> $W - W_n = o(n^{-\alpha})$ <u>if and only if</u>

------

1) In such statements is frequently understood a.s.

$$(1.2)\quad \int_y^\infty x[\log x - \log y]\, dF(x) = o([\log y]^{-\alpha}).$$

(iii) Let $\alpha > 0$. Then $\Sigma_{n=0}^{\infty} n^{\alpha-1}\{W - W_n\}$ converges if and only if $\mu_{\alpha+1} < \infty$, where

$$\mu_\beta = \int_0^\infty x[\log^+ x]^\beta\, dF(x)$$

[to get a feeling for (1.2), note that

$$(1.3)\quad \mu_{\alpha+1} < \infty \Rightarrow \int_y^\infty x \log x\, dF(x) = o([\log y]^{-\alpha}) \Rightarrow (1.2) \Rightarrow \mu_{\alpha+1-\varepsilon} < \infty\ \forall\varepsilon > 0$$

as is easily seen upon integration by parts]. 1.2 is a slight sharpening and extension of [1]. Results of similar form can be found in the theory of sums of i.i.d.r.v. $U_1, U_2, \ldots$ For example, letting $\mu = EU_1$, $\overline{U}_n = (U_1 + \ldots + U_n)/n - \mu$, it holds that

$$(1.4)\quad \overline{U}_n = o(n^{-1/q}) \Leftrightarrow E|U_1|^p < \infty \quad (1 < p < 2,\ 1/p + 1/q = 1)$$

$$(1.5)\quad \overline{U}_n = o([\log n]^{-\alpha}) \Leftrightarrow E|U_1|\ [\log^+|U_1|]^\alpha < \infty \quad (\alpha \geqq 0)$$

see [14], pg. 152-155, slightly extended. Together with

$$(1.6)\quad \sup_n W_n < \infty, \quad P(\inf_n W_n > 0) = P(W > 0) > 0$$

$$(1.7)\quad EW^p < \infty \Leftrightarrow \int_0^\infty x^p\, dF(x) < \infty \quad (p > 1)$$

$$(1.8)\quad EW[\log^+ W]^\alpha < \infty \Leftrightarrow \mu_{\alpha+1} < \infty$$

which holds assuming ($x \log x$), see 1.1 and [6], [9], (1.4) and (1.5) also provide a first motivation for 1.2, since conditioned upon $\mathcal{F}_n$ $W - W_n$ is distributed as $W_n \overline{U}_{Z_n}$ if we let $P(U \leqq u) = P(W \leqq u)$. To see this, let $W^{n,i}$ be the W-variable corresponding to the Galton-Watson process initiated by the ith individual alive at time n and note that

$$(1.9)\quad W - W_n = \frac{1}{m^n} \sum_{i=1}^{Z_n} \{W^{n,i} - 1\} = W_n \frac{1}{Z_n} \sum_{i=1}^{Z_n} \{W^{n,i} - 1\}\ .$$

Noting that $Z_n \simeq m^n$ and combining (1.4), (1.7), (1.9) leads precisely to part (i) of 1.2, while using instead (1.5), (1.8), (1.9) one is lead to expect the condition for $W - W_n = o(n^{-\alpha})$ to be $\mu_{\alpha+1} < \infty$, which is only slightly stronger that (1.2), cf. (1.3). However, in § 3 we sketch a different point of view on 1.2.

Also the technique in our proofs of 1.1, 1.2 relates to sums of independent r.v. As relevant background, we suggest to keep in mind Kolmogorov's three series criterion and the somewhat related standard proof of (1.4), (1.5), see [14], pg. 152-155. A common feature is here an approximation argument, which for example for the law of large numbers consists in studying

$$(1.10) \quad \sum_{n=0}^{\infty} \{\tilde{U}_n - E\tilde{U}_n\}/n = \sum_{n=0}^{\infty} \{\tilde{U}_n - \mu + EU_n I(|U_n| > n)\}/n,$$

where $\tilde{U}_n = U_n I(|U_n| \leq n)$, rather than the (non necessarily convergent) series $\Sigma\{U_n - \mu\}/n$. Adapting this idea to the Galton-Watson process, we study not the series (1.1) but instead a series

$$(1.11) \quad \sum_{n=0}^{\infty} \{\tilde{W}_{n+1} - E(\tilde{W}_{n+1} \mid F_n)\} = \sum_{n=0}^{\infty} \{\tilde{W}_{n+1} - W_n + R_n\}$$

defined in analogy by (1.10), that is, by

$$\tilde{W}_{n+1} = m^{-n-1} \sum_{i=1}^{Z_n} X_{n,i} \; I(X_{n,i} \leq c_n) ,$$

$$R_n = E(W_n - \tilde{W}_{n+1} \mid F_n) = E(W_{n+1} - \tilde{W}_{n+1} \mid F_n) = m^{-1} W_n \int_{c_n}^{\infty} x \, dF(x) .$$

By definition, (1.11) is again a martingale series. As a common feature in the proofs enters a routine calculation of three series similar to those of Kolmogorov,

$$(1.12) \quad \sum_{n=0}^{\infty} P(\tilde{W}_{n+1} \neq W_{n+1}), \quad \sum_{n=0}^{\infty} ER_n, \quad \sum_{n=0}^{\infty} \mathrm{Var}\{\tilde{W}_{n+1} - W_n + R_n\}$$

It is here where the moment conditions on F come in, but the calculations in (1.12) alone does not prove the results. Additional ideas varying from case to case are required to complete the proofs.

There are numerous ways of varying the basic model and when developing techniques for dealing with the Galton-Watson process, it is important that these can be used in more general branching processes. The adaptation to processes with several (even infinitely many) types has already been presented as part of [2] and we treat here in § 5 age-dependent processes. As example we have chosen the Bellman-Harris process ([12], Ch. 6) and give a full treatment of the limit theory. It turns out that our proof of the analogue of 1.1 with some minor modifications provide one of the basic lemmas needed when treating the further a.s. convergence results on the distribution of the population according to ages, see e.g. [12], [7]. We have added sufficient material to make the exposition totally self contained and borrow here some ideas from [4], [7] as well as [2].

Finally, § 4 is devoted to a remark on results of type (1.7), (1.8). The proofs in the literature, see e.g. [6], [9], are in part both deep and laborious and we sketch a different approach based on moment inequalities for sums of independent r.v. rather than expansions of Laplace transforms.

2. THE x log x CONDITION

Our first example on the use of the martingales $\Sigma\{\widetilde{W}_{n+1} - W_n + R_n\}$ defined in § 1 is to give the proof of 1.1. We let $c_n = m^n$ and the series in (1.12) are then computed the following way:

$$(2.1) \quad \sum_{n=0}^{\infty} P(W_{n+1} \neq \widetilde{W}_{n+1}) =$$

$$\sum_{n=0}^{\infty} EP(X_{n,i} > m^n \text{ for some } i = 1,\dots,Z_n \mid F_n) \leq$$

$$\sum_{n=0}^{\infty} EZ_n \int_{m^n}^{\infty} dF(x) = \int_0^{\infty} \sum_{n=0}^{\infty} m^n I(x > m^n)\, dF(x) = \int_0^{\infty} 0(x)\, dF(x)$$

$$(2.2)\quad \sum_{n=0}^{\infty} ER_n = m^{-1} \sum_{n=0}^{\infty} \int_{m^n}^{\infty} x\, dF(x) =$$

$$m^{-1} \int_0^{\infty} x\Big(\sum_{n=0}^{\infty} I(x > m^n)\Big)\, dF(x) = \int_0^{\infty} x\, 0(\log^+ x)\, dF(x)$$

$$(2.3)\quad \sum_{n=0}^{\infty} \mathrm{Var}\{\widetilde{W}_{n+1} - W_n + R_n\} = \sum_{n=0}^{\infty} E\,\mathrm{Var}(\widetilde{W}_{n+1} \mid F_n) =$$

$$\sum_{n=0}^{\infty} E\, m^{-2n-2} Z_n\, \mathrm{Var}[X_{1,1}\, I(X_{1,1} \leqq m^n)] \leqq$$

$$m^{-2} \sum_{n=0}^{\infty} m^{-n} \int_0^{m^n} x^2\, dF(x) = m^{-2} \int_0^{\infty} x^2\Big(\sum_{n=0}^{\infty} m^{-n}\, I(x \leqq m^n)\Big) dF(x) =$$

$$\int_0^{\infty} x^2\, 0(x^{-1})\, dF(x) = \int_0^{\infty} 0(x)\, dF(x).$$

From (2.3) and the convergence theorem for L^2-bounded martingales, we obtain

<u>2.1 Lemma</u> <u>Without any moment conditions on</u> F <u>beyond</u> $1 < m < \infty$, $\Sigma\{\widetilde{W}_{n+1} - W_n + R_n\}$ <u>converges a.s. and in</u> L^1.

$EW \leqq 1$ is immidiate form Fatous lemma. To prove the converse, assuming (x log x), we let $N \to \infty$ in the inequality

$$EW = E\Big(W_0 + \sum_{n=0}^{N} \{W_{n+1} - W_n\} + \sum_{n=N+1}^{\infty} \{W_{n+1} - W_n\}\Big)$$

$$\geqq 1 + 0 + E\Big(\sum_{n=N+1}^{\infty} \{\widetilde{W}_{n+1} - W_n\}\Big)$$

which is obvious from $W_{n+1} \geqq \widetilde{W}_{n+1}$. We then only have to prove the L^1-convergence of $\Sigma\{\widetilde{W}_{n+1} - W_n\}$ which in view of the lemma is equivalent to that of ΣR_n. Since $R_n \geqq 0$, it suffices that $\Sigma ER_n < \infty$, which follows from (2.2).

To prove that $W = 0$ if (x log x) fails, we first note that the existence of $W = \lim_n W_n$ implies the a.s. convergence of the telescoping series $\Sigma\{W_{n+1} - W_n\}$ and therefore that of $\Sigma\{\widetilde{W}_{n+1} - W_n\}$, since $\widetilde{W}_{n+1} = W_{n+1}$ for n large by (2.1) and the Borel-Cantelli lemma.

Combining this with Lemma 2.1, we have a.s. convergence of ΣR_n. Now let $\underline{W} = \inf_n W_n$ and note that $\{W > 0\} = \{\underline{W} > 0\}$. If (x log x) fails, $\Sigma\int_{m^n}^{\infty} x\, dF(x) = \infty$ as in (2.2) and $P(W > 0) = 0$ follows from

$$\infty > \sum_{n=0}^{\infty} R_n \geq m^{-1}\, \underline{W} \int_{m^n}^{\infty} x\, dF(x) = \infty \quad \text{on } \{\underline{W} > 0\}\ .$$

3. CONVERGENCE RATES

We next consider the proof of 1.2. To estimate $W - W_N$ we write $W - W_N = \Sigma_N^{\infty} \alpha_n$ where $\alpha_n = W_{n+1} - W_n$ and use

<u>3.1 Lemma</u> <u>Let</u> $\{\alpha_n\},\{\beta_n\}$ <u>be series of real numbers such that</u> $0 < \beta_n \uparrow \infty$. <u>Then</u>

$$(3.1)\quad \sum_{n=0}^{\infty} \alpha_n \beta_n \text{ converges} \Rightarrow \sum_{n=N}^{\infty} \alpha_n = o(1/\beta_N)$$

Obviously (3.1) is analogous to Kronecker's lemma, which is used in the proof of (1.4), (1.5) and states that under the assumption of (3.1) it holds that

$$(3.2)\quad \sum_{n=0}^{\infty} \alpha_n/\beta_n \text{ converges} \Rightarrow \sum_{n=0}^{N} \alpha_n = o(\beta_N)$$

In fact, both (3.1) and (3.2) are immediate from Abel's lemma, [10] pg. 54.

We first consider part (ii) of (1.2), the proof of which is particularly well suited to demonstrate the ideas. We let $\beta_n = n^{\alpha}$ and instead of studying $\Sigma\alpha_n \beta_n = \Sigma n^{\alpha}\{W_{n+1} - W_n\}$, we approximate by $\Sigma n^{\alpha}\{\widetilde{W}_{n+1} - W_n + R_n\}$ defined as in § 1 with $c_n = m^n/n^{\alpha}$. Calculations similar to (2.1), (2.3) yields

$$\sum_{n=0}^{\infty} P(\widetilde{W}_{n+1} \neq W_{n+1}) \simeq \sum_{n=0}^{\infty} \mathrm{Var}[n^{\alpha}\{\widetilde{W}_{n+1} - W_n + R_n\}] \simeq \mu_{\alpha}$$

and as in § 2, we have immediately

<u>3.2 Lemma</u> <u>Let</u> $\alpha > 0$, $c_n = m^n/n^{\alpha}$ <u>and suppose</u> $\mu_{\alpha} < \infty$. <u>Then</u>

$\Sigma n^{\alpha}\{\tilde{W}_{n+1} - W_n + R_n\}$, $\Sigma n^{\alpha}\{W_{n+1} - W_n + R_n\}$ converges a.s.

Proof of (ii). Suppose first $\mu_{\alpha} < \infty$ (which is substantially weaker than (1.2), cf. (1.3)). Combining 3.1 and 3.2 yields

$$o(N^{-\alpha}) = \sum_{n=N}^{\infty} \{W_{n+1} - W_n + R_n\} = W - W_N + \sum_{n=N}^{\infty} R_n \ .$$

Therefore $W - W_N = o(N^{-\alpha})$ is equivalent to

$$o(N^{-\alpha}) = \sum_{n=N}^{\infty} R_n = \sum_{n=N}^{\infty} m^{-1} W_n \int_{m^n/n^{\alpha}}^{\infty} x \, dF(x)$$

or, appealing to (1.6), to

$$(3.3) \quad o(N^{-\alpha}) = \sum_{n=N}^{\infty} \int_{m^n/n^{\alpha}}^{\infty} x \, dF(x) \ .$$

Define $y_n = m^n/n^{\alpha}$, $N(x) = \sup\{n : y_n \leqq x\}$. Then (3.3) can be rewritten as

$$(3.4) \quad o([\log y_N]^{-\alpha}) = \int_{y_N}^{\infty} x(N(x) - N) \, dF(x), \quad N \to \infty.$$

Apparently (3.4) is weaker than

$$(3.5) \quad o([\log y]^{-\alpha}) = \int_{y}^{\infty} x(N(x) - N(y)) \, dF(x), \quad y \to \infty$$

but if (3.4) holds, so does (3.5) since for $y_N \leqq y < y_{N+1}$ then

$$\int_{y}^{\infty} x(N(x) - N(y)) \, dF(x) \leqq \int_{y_N}^{\infty} x(N(x) - N) \, dF(x) =$$

$$o([\log y_N]^{-\alpha}) = o([\log y]^{-\alpha}).$$

Now from the definition of N(x) it can be verified that

$$N(x) = \frac{\log x}{\log m} + \frac{\alpha}{\log m} \log \log x + O(1) \ .$$

As $x, y \to \infty$, the meanvalue theorem for the log yields $\log \log x - \log \log y = o(\log x - \log y)$ so that the right-hand side of (3.5) is

$$\int_{y}^{\infty} x(\log x - \log y)\left(\frac{1}{\log m} + o(1)\right) dF(x) + \int_{y}^{\infty} x \, O(1) \, dF(x) \ .$$

Since $\mu_\alpha < \infty$, the last term is $o([\log y]^{-\alpha})$ and therefore conditions (3.5) and (1.2) are equivalent, completing the proof when $\mu_\alpha < \infty$.

Suppose next $\mu_\alpha = \infty$. Then by (1.3), certainly (1.2) fails and we have to prove that $W - W_n = o(n^{-\alpha})$ must fail too. Since we assume $(x \log x)$, we can find β such that $1 \leqq \beta < \alpha$ and that $\mu_\beta < \infty$, $\mu_{\beta+1/2} = \infty$. Then from (1.3) and the first part of this proof it follows that $W - W_n = o(n^{-\beta})$ fails and the proof is complete since $\beta < \alpha$.□

<u>Proof</u> of (iii). Let $\beta_n = \Sigma_1^n k^{\alpha-1}$. Then

$$\sum_{n=1}^{N} \alpha_n \beta_n = \sum_{k=1}^{N} k^{\alpha-1} \sum_{n=k}^{\infty} \alpha_n - \beta_N \sum_{n=N+1}^{\infty} \alpha_n$$

and from (3.1) it follows by letting $N \to \infty$ that

(3.6) $\sum_{n=1}^{\infty} \alpha_n \beta_n$ converges $\Rightarrow$ $\sum_{k=1}^{\infty} k^{\alpha-1} \sum_{n=k}^{\infty} \alpha_n$ converges .

Let $\widetilde{W}_{n+1}$, R_n be defined as above with $c_n = m^n/n^\alpha$. Using $\beta_n \simeq n^\alpha$ one obtains

$$\sum_{n=0}^{\infty} P(\widetilde{W}_{n+1} \neq W_{n+1}) \simeq \sum_{n=0}^{\infty} \operatorname{Var}\{\beta_n\{W_{n+1} - W_n + R_n\}\} \simeq \mu_\alpha .$$

Thus if $\mu_\alpha < \infty$, $\Sigma\, \beta_n\{W_{n+1} - W_n + R_n\}$ converges a.s. and from (3.6), we get the a.s. convergence of

$$\sum_{k=0}^{\infty} k^{\alpha-1} \sum_{n=k}^{\infty} \{W_{n+1} - W_n + R_n\} = \sum_{k=0}^{\infty} k^{\alpha-1}\{W - W_k + \sum_{n=k}^{\infty} R_n\} .$$

Thus the convergence of $\Sigma\, k^{\alpha-1}\{W - W_k\}$ is equivalent to that of

$$\sum_{k=0}^{\infty} k^{\alpha-1} \sum_{n=k}^{\infty} R_n = m^{-1} \sum_{k=0}^{\infty} k^{\alpha-1} \sum_{n=k}^{\infty} W_n \int_{m^n/n}^{\infty} x\, dF(x)$$

or, appealing to (1.6), to that of

$$\sum_{k=0}^{\infty} k^{\alpha-1} \sum_{n=k}^{\infty} \int_{m^n/n}^{\infty} x\, dF(x) = \sum_{n=0}^{\infty} \beta_n \int_{m^n/n}^{\infty} x\, dF(x) .$$

Using $\beta_n \simeq n^\alpha$, this precisely reduces to $\mu_{\alpha+1} < \infty$ and the proof is complete when $\mu_\alpha < \infty$.

If $\mu_\alpha = \infty$, then of course $\mu_{\alpha+1} = \infty$. Assuming $(x \log x)$ we can choose β, $1 \leq \beta < \alpha$, such that $\mu_\beta < \infty$, $\mu_{\beta+1} = \infty$. Then the first part of the proof excludes the convergence of $\Sigma\, n^{\beta-1}\{W - W_n\}$ and Abel's criterion ([10], pg. 48) that of $\Sigma\, n^{\alpha-1}\{W - W_n\}$.□

<u>Proof</u> of (i). We let $\beta_n = m^{n/q}$, $c_n = m^{n/p}$. Then

$$(3.7) \quad \sum_{n=0}^{\infty} P(\tilde{W}_{n+1} \neq W_{n+1}) \simeq \sum_{n=0}^{\infty} \mathrm{Var}[m^{n/q}\{\tilde{W}_{n+1} - W_n + R_n\}] \simeq \int_0^\infty x^P \, dF(x).$$

Assuming the right-hand side to be finite, we have a.s. convergence of $m^{n/q}\{W_{n+1} - W_n + R_n\}$, $m^{n/q}\{W_{n+1} - W_n + R_n\}$ and (3.1) gives

$$o(m^{-N/q}) = W - W_n + \sum_{n=N}^{\infty} R_n = W - W_N + m^{-1} \sum_{n=N}^{\infty} W_n \int_{m^{n/q}}^{\infty} x \, dF(x) .$$

But the last term is $o(m^{-N/q})$, since

$$\sum_{n=0}^{\infty} m^{n/q} \int_{m^{n/p}}^{\infty} x \, dF(x) = \int_0^\infty O(x^P) \, dF(x) < \infty$$

and it follows that $W - W_N = o(m^{-N/q})$, proving one way of the result.

For the converse, the method in the proofs of part (ii), (iii) and in § 2 does not apply, because the condition for convergence in (3.7) is not weaker than that for the result. Our proof is here totally different and we proceed by reducing the necessity problem for the Galton-Watson process to that of sums of i.i.d.r.v., cf. (1.4).

Suppose $W - W_N = o(m^{-N/q})$. In particular, $W_{n+1} - W_n = o(m^{-n/q})$ so that on $\{W > 0\}$

$$(3.8) \quad Z_n^{-1/p} \sum_{i=1}^{Z_n} \{X_{n,i} - m\} = W_n^{-1/p}\, m^{n/q} \{W_{n+1} - W_n\} \to 0 .$$

Let the r.v. U_n in (1.4) be distributed as $X_{n,i} - m$, let $q(n,\varepsilon) = P(n^{1/q}|\overline{U}_n| > \varepsilon)$ and let $U_1^c, U_2^c, \ldots$ be independent and follow the symmetrized distribution of U_n, that is, the distribution of $X_{n,1} - X_{n,2}$. Define $\overline{U}_n^c$, $q^c(n,\varepsilon)$ the obvious way. It is then well-known that U_n^c has p^{th} moment if and only if F has so, so by

(1.4) it suffices to prove that $\overline{U}_n^c = o(n^{-1/q})$. By the conditional Borel-Cantelli lemma and (3.8), we have on $\{W > 0\}$

$$\sum_{n=0}^{\infty} q(Z_n,\varepsilon) = \sum_{n=0}^{\infty} P(|Z_n^{-1/p} \sum_{i=1}^{Z_n} \{X_{n,i} - m\}| > \varepsilon \mid F_n) < \infty$$

and therefore also by a standard inequality $\Sigma\, q^c(Z_n,2\varepsilon) \leq 2\, \Sigma\, q(Z_n,\varepsilon) < \infty$. Now pick a numerical sequence $\{k(n)\}$ of integers of the form $k(n) = Z_n(\omega)$, where ω belongs to the set of positive probability where $W > 0$, $\Sigma\, q^c(Z_n,\varepsilon) < \infty$ for all rational (and therefore all) $\varepsilon > 0$. Then

$$(3.9) \quad \sum_{n=0}^{\infty} P(k(n)^{1/q} \, |\overline{U}_{k(n)}^c| > \varepsilon) = \sum_{n=0}^{\infty} q^c(k(n),\varepsilon) < \infty$$

implying $\overline{U}_{k(n)}^c = o(k(n)^{-1/q})$. Define

$$M_n = k(n)^{-1/p} \max_{1 \leq i \leq k(n)} |U_1^c + \ldots + U_i^c| .$$

By Levy's inequality and (3.7)

$$\sum_{n=0}^{\infty} P(M_n > \varepsilon) \leq 2 \sum_{n=0}^{\infty} P(k(n)^{-1/p} |U_1^c + \ldots + U_{k(n)}^c| > \varepsilon) =$$

$$2 \sum_{n=0}^{\infty} q^c(k(n),\varepsilon) < \infty$$

so that $M_n \to 0$. One checks readily that when $k(n) \leq i \leq k(n+1)$,

$$i^{1/q} \, |\overline{U}_i^c| \leq k(n)^{1/q} \, |\overline{U}_{k(n)}^c| + 2 \left(\frac{k(n+1)}{k(n)}\right)^{1/p} M_{n+1}$$

and $\overline{U}_i^c = o(i^{-1/q})$, $i \to \infty$, follows since $k(n+1)/k(n) \to m$ (in particular, the sequence $k(n)$ is ultimately increasing).□

We conclude by some remarks on the relation of 1.2 to sums of independent r.v. It is possible to exploit the motivation for 1.2 given in § 1 somewhat further by using (1.7), (1.9) to prove

$$\sum_{n=0}^{\infty} P(m^{n/q} \, |W - W_n| > \varepsilon \mid F_n) < \infty \quad \text{if} \quad \int_0^{\infty} x^p \, dF(x) < \infty$$

and thus one half of (i). Similarly, (1.8) and (1.9) combine to give $W - W_N = o(N^{-\alpha})$ if $\mu_{\alpha+1} < \infty$. However, the full strength of 1.2

does not seem to follow this way and as is apparent from the proofs, we exploit the structure of $W - W_n$ as the tail sum of (1.1) rather than (1.9). Also, as remarked earlier not all results are the perfect analogues of (1.4), (1.5) to be expected from (1.7), (1.8), (1.9). Instead we state the following result on sums of independent r.v., whose form and proof is more similar.

3.3 Theorem Let $U_1, U_2, \ldots$ be i.i.d. with common distribution G. Then (from Kolmogorov's three series criterion) $\Sigma\, U_n/n$ converges if and only if

$$(3.10) \quad \int_{-\infty}^{\infty} |x|\, dG(x) < \infty, \quad \sum_{n=0}^{\infty} \int x\, I(|x| \leq n)\, dG(x) \quad \text{converges}$$

Suppose (3.10) holds and that $EU_n = 0$. Then:

(i) $\Sigma_N^{\infty} U_n/n = o(N^{-1/q})$ if and only if

$$\int_{-\infty}^{\infty} |x|^p\, dG(x) < \infty \quad (1 < p < 2,\ 1/p + 1/q = 1).$$

(ii) For $\alpha > 0$, $\Sigma_N^{\infty} U_n/n = o([\log N]^{-\alpha})$ if and only if

$$\mu_\alpha = \int_{-\infty}^{\infty} |x|\ |\log^+ x|^{\alpha}\, dG(x) < \infty,$$

(3.11)

$$\sum_{n=N}^{\infty} \int x\, I(|x| > n/(\log n)^{\alpha})\, dG(x) = o([\log N]^{-\alpha}).$$

There are, of course, similar results for other weights than n^{-1} and also part (iii) of 1.2 has a counterpart. The conditions (3.10), (3.11) can not be expressed in terms of the μ_α in the same way as in (1.3). For example, if G is symmetric, (3.11) reduces to $\mu_\alpha < \infty$, while if G is concentrated on $[a, \infty)$ for some $a > -\infty$, then (3.11) reduces to (1.2) (with F replaced with G) and (1.3) holds.

4. A REMARK ON THE MOMENTS OF W

We recall the results (1.7), (1.8) concerning the relation between the moments in the offspring distribution F and those of W.

The aim of the present section is to sketch an approach different from that of [6], [9] to results of this type.

As set-up, we choose to consider moments of the form $EW^{\nu}f(W)$, where ν is an integer and f a suitable function satisfying $f(x) = o(x)$, $x \to \infty$, for example $f(x) = x^{\alpha}$, $0 < \alpha < 1$. A detailed treatment is given only for the case $\nu = 1$, which is of particular importance and suffices to demonstrate the ideas.

<u>4.1 Lemma</u> <u>Let</u> $f: [0,\infty) \to [0,\infty)$ <u>be concave and let</u> $S = X_1 + \ldots + X_N$ <u>be a sum of</u> N <u>independent r.v.</u> $X_i \geqq 0$. <u>Then</u>

$$(4.1)\quad ESf(S) \leqq ESf(ES) + \sum_{i=1}^{N} EX_i f(X_i)\,.$$

<u>Proof</u> The assumptions on f imply subadditivity, $f(a+b) \leqq f(a) + f(b)$, $a,b \geqq 0$. Thus

$$ESf(S) = \sum_{i=1}^{n} EX_i f(S) \leqq \sum_{i=1}^{n} \{EX_i f(\sum_{j \neq i} X_j) + EX_i f(X_i)\}$$

$$\leqq ESf(ES) + \sum_{i=1}^{n} EX_i f(X_i)\ ,$$

since by Jensen's inequality

$$EX_i f(\sum_{j \neq i} X_j) = EX_i Ef(\sum_{j \neq i} X_j) \leqq EX_i f(E\sum_{j \neq i} X_j) \leqq EX_i f(ES).\square$$

Letting $N = Z_n$, $X_i = X_{n,i}/m^{n+1}$, $S = m^{-n-1}\sum_{i=1}^{Z_n} X_{n,i} = W_{n+1}$ yields

$$E(W_{n+1}\, f(W_{n+1}) \mid \mathcal{F}_n) \leqq$$

$$E(W_{n+1} \mid \mathcal{F}_n)\, f(E(W_{n+1} \mid \mathcal{F}_n)) + \sum_{i=1}^{Z_n} E\left(\frac{X_{n,i}}{m^{n+1}}\, f\left(\frac{X_{n,i}}{m^{n+1}}\right) \middle|\ \mathcal{F}_n\right) =$$

$$W_n f(W_n) + m^{-1} W_n \int_0^{\infty} x\, f\left(\frac{x}{m^{n+1}}\right) dF(x)$$

and it follows that

$$E\,W f(W) \leqq \underline{\lim}\, E\, W_{N+1} f(W_{N+1}) =$$

$$\underline{\lim}\,\{f(1) + \sum_{n=0}^{N} \{EW_{n+1}f(W_{n+1}) - EW_n f(W_n)\}\} \leqq$$

$$f(1) + \sum_{n=0}^{\infty} m^{-1}\ EW_n \int_0^{\infty} x\, f\left(\frac{x}{m^{n+1}}\right) dF(x) =$$

$$f(1) + m^{-1} \int_0^{\infty} x \sum_{n=0}^{\infty} f\left(\frac{x}{m^{n+1}}\right) dF(x) .$$

4.2 Example Let $1 < p < 2$, $f(x) = x^{p-1}$. Computation of $\Sigma\, f(x/m^{n+1})$ and inserting yields

$$(4.2)\quad EW^p \leqq 1 + \frac{\int_0^{\infty} x^p\, dF(x)}{m^p - m}, \quad 1 < p < 2.$$

In particular, $EW^p < \infty$ if the pth moment in the offspring distribution is finite. The converse is immediate assuming $(x \log x)$, since then by convexity

$$\int_0^{\infty} x^p\, dF(x) = m^p\ EW_1^p = m^p\ E(E(W \mid F_1))^p \leqq m^p\ EW^p.$$

4.3 Example In (1.8), $[\log^+ x]^\alpha$ does not satisfy the assumption on $f(x)$, but so does

$$f(x) = \begin{cases} c_1 x & 0 \leqq x \leqq x_0 \\ [\log^+ x]^\alpha + c_2 & x_0 \leqq x < \infty \end{cases}$$

if we chose first $x_0 > 1$ such that $d^2/dx^2\ (\log x)^\alpha < 0$ when $x \geqq x_0$ and let

$$c_1 = \frac{d}{dx}(\log x)^\alpha \Big|_{x=x_0}, \quad c_2 = c_1\, x_0 - (\log x_0)^\alpha$$

$\Leftarrow$ in (1.8) follows at once, since one easily checks $\Sigma\, f(x/m^{n+1}) = O([\log^+ x]^{\alpha+1})$.

We shall not here further work out the approach. Some problems, in particular to prove $\Rightarrow$ in (1.8) seems to require additional ideas, while others are immediate. For example, the method works in the multitype or age-dependent case with a mere change of notation by studying the one-dimensional martingale functionals of the process. Also moments of order higher than the second can be treated. We state here the following inequality, which is valid for $\nu = 1,2,\ldots$ under the hypothesis of 4.1:

$$(4.3) \quad ES^{\nu}f(S) \leqq ES^{\nu}f(ES) + \sum_{\mu=1}^{\nu} \binom{\nu}{\mu} ES^{\nu-\mu} \sum_{i=1}^{N} EX_i^{\mu}f(X_i).$$

5. THE LIMIT THEORY OF THE SUPERCRITICAL BELLMAN-HARRIS PROCESS

The model is the following. All individuals have lifelengths governed by a distribution G on $(0,\infty)$. At the time of death of the parent a random number of children are born according to the offspring distribution F. The lifelength and number of children of any particular individual are independent, and all individuals evolve independently of each other.

For questions of existence and construction, we refer to [12]. As remarked at a number of occasions in the literature (going back at least to [12]), the process is most naturally considered as a Markovian multitype process identifying types with ages. Accordingly, we define the state Z_t of the process at time t not as the number n of individuals alive, but as the collection $Z_t = \langle x_1,\ldots,x_n \rangle$ of their ages. By averaging Z_t with various η belonging to the set B of bounded measurable functions on $[0,\infty)$, we obtain a number of functionals useful in the study of the process, defined by $Z_t[\eta] = 0$ if the population is extinct at time t and by

$$Z_t[\eta] = \eta(x_1) + \ldots + \eta(x_n) \quad \text{if} \quad Z_t = \langle x_1,\ldots,x_n \rangle \quad .$$

For example, $|Z_t| = Z_t[1]$ is the total population size. Also, if we think of Z_t as a (random) measure on $[0,\infty)$, then simply

$$Z_t[\eta] = \int_0^{\infty} \eta(x)\, Z_t[dx].$$

Specific assumptions on Z_0 are usually not relevant but, whenever needed, P^x, E^x etc. refer to the case $Z_0 = \langle x \rangle$. We throughout consider the supercritical case

$$1 < m = \int_0^\infty x\, dF(x) < \infty$$

and assume as usual that G is non-lattice with $G(0) = 0$. Define $\alpha > 0$ as the (unique) root of

$$m \int_0^\infty e^{-\alpha x}\, dG(x) = 1$$

and let

$$A[dx] = e^{-\alpha x}(1 - G(x))\, dx \Big/ \int_0^\infty e^{-\alpha y}(1 - G(y))\, dy$$

$$G^x(t) = (G(x+t) - G(x)) \,/\, (1 - G(x))$$

$$n_1 = \int_0^\infty y\, e^{-\alpha y}\, dG(y) \Big/ \int_0^\infty e^{-\alpha y}(1 - G(y))\, dy$$

$$V(x) = n_1^{-1} \int_0^\infty e^{-\alpha y}\, dG^x(y)$$ 1)

$$M_t\eta(x) = E^x Z_t[\eta]\,, \quad \mu M_t[\eta] = \int_0^\infty M_t\eta(x)\, \mu[dx]\,.$$

It is then readily checked that $\{M_t\}_{t \geq 0}$ is a semigroup acting to the right on the set B of bounded Borel-measurable functions η on $[0,\infty)$ and to the left on the set of bounded measures μ on $[0,\infty)$. Furthermore:

5.1 Lemma A,V *are eigenfunctions of* M_t *corresponding to the eigenvalue* $e^{\alpha t}$, *i.e.*

(5.1) $AM_t = e^{\alpha t} A\,, \quad M_t V = e^{\alpha t} V\,.$

Furthermore for any $\eta \in B$ *such that* $e^{-\alpha x}(1 - G(x))\eta(x)$ *is directly Riemann integrable* (cf. [11], pg. 361 - 362)

----

1) There is some ambiguity in the literature concerning the normalization of V. The present choice ensures $A[1] = A[V] = 1$, $V(0) = (m\, n_1)^{-1}$.

$$(5.2)\quad \sup_{0\le x<\infty} |e^{-\alpha t} M_t \eta(x) - V(x)\, A\,[\eta]| \to 0 \ ,\ t\to\infty\ .$$

The class of η's satisfying the assumptions for (5.2) is rather extensive and contains e.g. for all $0\le a\le b\le\infty$ $\eta(x) = I(a\le x<b)$. Thus (5.2) states that in the mean the population at time t is asymptotically composed like the measure $e^{\alpha t}\, V(x)\, A$, where x is the age of the ancestor, and for this reason and (5.1), A is usually called the stable age-distribution, V the reproductive value and α the Malthusian parameter, cf. [12].

5.2 Remark Suppose the ancestor is of age x and let λ be the time of his death. Then from time λ on the process evolves like the sum of N independent processes with ancestors of age zero, N chosen at random according to F. In particular, P^x depends only on x through G^x. This explains somewhat further the role of λ and V, since $V(x) = n_1^{-1}\, E^x e^{-\alpha\lambda}$.

Proof of 5.1. We first prove (5.2). Let for some fixed η satisfying the assumptions $K^x(t) = E^x\, Z_t[\eta]$, $\tilde{K}(t) = e^{-\alpha t}\, K^0(t)$, $d\tilde{G}(x) = m e^{-\alpha x}\, dG(x)$. Appealing to 5.2,

$$(5.3)\quad K^x(t) = E^x Z_t[\eta]\, I(\lambda>t) + E^x(|Z_\lambda|\ K^0(t-\lambda)\ I(\lambda\le t))$$

$$= \quad (x+t)\ \frac{1-G(x+t)}{1-G(x)} \ + \int_0^t m\, K^0(t-u)\ \frac{dG(x+u)}{1-G(x)}\quad .$$

Letting $x=0$ and multiplying by $e^{-\alpha t}$ gives

$$\tilde{K}(t) = e^{-\alpha t}\ \eta(t)(1-G(t)) + \int_0^t \tilde{K}(t-u)\ d\tilde{G}(u)\ .$$

The choice of α ensures that $\tilde{G}$ is a probability measure so that by the renewal theorem

$$\lim_{t\to\infty} \tilde{K}(t) = \frac{\int_0^\infty e^{-\alpha t}\ \eta(t)(1-G(t))\ dt}{\int_0^\infty t\ d\tilde{G}(t)} = \frac{A[\eta]}{m\, n_1} = V(0)\, A\,[\eta]\ .$$

Inserting in (5.3), (5.2) follows after some elementary estimates.

(5.1) is an easy consequence of (5.2). For example integrating (5.2) w.r.t. A yields $e^{-\alpha t} A M_t \eta \to A[\eta]$ for all η satisfying the assumptions for (5.2) and therefore by weak continuity for all a.e. [1] continuous η, cf. [8]. It is not difficult to see, that if $\eta \in B$ is continuous, then $M_s \eta$ is a.e. continuous. Therefore

$$A\, M_s[\eta] = A[M_s \eta] = \lim_{t\to\infty} e^{-\alpha t} A\, M_t\, M_s\, \eta$$

$$= \lim_{t\to\infty} e^{\alpha s}\, e^{-\alpha(t+s)}\, A\, M_{t+s}\eta = e^{\alpha s}\, A\,[\eta]$$

and $A\, M_s = e^{\alpha s} A$ follows. $M_s\, V = e^{\alpha s}\, V$ is proved in a similar manner. □

Let F_t be the σ-algebra containing all relevant information on the process up to time t. From (5.2), we get

$$E(Z_{t+s}[V] \mid F_t) = Z_t M_s[V] = e^{\alpha s}\, Z_t[V]$$

and it follows that $\{W_t\}_{t \geqq 0}$, where $W_t = e^{-\alpha t}\, Z_t[V]$, is a non-negative martingale w.r.t. $\{F_t\}_{t\geqq 0}$. Thus $W = \lim_t W_t$ exists a.s. and the main result on the limiting behaviour of the process is the following, the proof of which occupies the rest of this section:

<u>5.3 Theorem</u> $E^x W = V(x)$, $x \geqq 0$, <u>if and only if</u>

(x log x) $\int_0^\infty x \log x \, dF(x) < \infty$

<u>while</u> $P^x(W = 0) = 1, \forall x \geqq 0$, <u>otherwise. Furthermore, for any</u> $\eta \in B$ <u>continuous a.e.</u>,

(5.4) $\lim_{t\to\infty} e^{-\alpha t}\, Z_t[\eta] = W\, A[\eta]$.

Compared with the Galton-Watson process, the complications occur from the fact that the different lines of descent still evolve independently, but no longer according to the same law. That is, if $Y \geqq 0$ is some functional of the process, $P^x(Y > y)$ depends on x.

----

1) Since A has a density, continuity a.e. on the essential span of F w.r.t. A or w.r.t. Lebesgue measure are the same concept.

We work here as in [7] with the assumption

(5.5) $P^x(Y > y) \leq 1 - H(y)$

where H is some distribution on $[0,\infty)$ independent of x. The reduction to (5.5) follows essentially from 5.4 below. In the proof, we adapt as everywhere in the following without further explanation the convention, that $Y^{t,i}$ <u>denotes the corresponding functional of the line of descent initiated by the ith individual alive at time t.</u>

<u>5.4 Lemma</u> <u>Let</u> $t > 0$ <u>and let</u> $Y = Y_t$ <u>be the total number of individuals which ever lived up to time</u> t. <u>Then</u> (5.5) <u>holds, where</u> H <u>may be taken with finite mean and satisfying</u>

$$(5.6)\quad (x \log x) \Rightarrow \int_0^\infty x \log x \, dH(x) < \infty .$$

In the proof, we need

<u>5.5 Lemma</u> <u>Let</u> $N, U_1, U_2, \ldots$ <u>be independent and non-negative with</u> N <u>integer-valued and</u> $U_1, U_2, \ldots$ <u>i.i.d. and let</u> $S = 1 + U_1 + \ldots + U_N$. <u>Define</u>

$$\log^* x = \begin{cases} x/e & 0 \leqq x \leqq e \\ \log x & x \geqq e \end{cases},$$

$\mu = E\, U_1 \log^* U_1$. <u>Then there exist constants</u> $c(\nu) < \infty$, $\nu \geqq 0$ (dependent on the distribution of N) <u>such that if</u> $\mu < \infty$, $E\, U_1 \leqq \nu$, $E\, N \log^* N < \infty$ <u>then</u> $E\, S \log^* S \leqq c(\nu) + \mu\, E\, N$.

<u>Proof</u> Since log* satisfies the assumptions of 4.1, we have

$$E(S \log^* S \mid N) \leqq E(S \mid N) \log^*(E(S \mid N)) + 1 \log^* 1 + N\mu$$

so we have only to let $c(\nu) = E(1 + N\nu) \log^*(1 + N\nu) + e^{-1}$. □

<u>Proof</u> of 5.4 Let $N, U_1, U_2, \ldots$ be independent with $P(N \leqq x) = F(x)$, $P(U_i \leqq u) = P^0(Y \leqq u)$ and let $S = 1 + \Sigma_1^N U_i$, $H(y) = P(S \leqq y)$. Letting N be the number of children born at time λ we have, appealing to 5.2,

$$P^x(Y_t > y) \leqq P^x(Y_{\lambda+t} > y) \leqq P(S > y) = 1 - H(y)$$

and we have to prove $\int_0^\infty x\,dH(x) < \infty$ and (5.6). We treat only the latter and more complicated case, which obviously is equivalent to $E\,S\log^*S < \infty$, or appealing to 5.5, to $\infty > E\,U_1 \log^* U_1 = E^0 Y_t \log^* Y_t = \mu(t)$ (say). Let $A_n(t)$ be the event that at most n deaths occur before time t. Obviously,

$$(5.7)\quad Y_t\ I(A_{n+1}(t)) \leq 1 + \sum_{i=1}^{|Z_\lambda|} Y_{t-\lambda}^{\lambda,i}\ I(A_n(t))$$

where for convenience $Y_s = 0$, $s < 0$. Define

$$\mu_n(t,\lambda) = E^0(Y_t \log^* Y_t\ I(A_n) \mid \lambda),\quad \mu_n(t) = E^0\ \mu_n(t,\lambda)$$

Letting $\nu = E^0\ Y_T$ where $T > t$ is fixed in 5.5 and using (5.7) gives

$$\mu_{n+1}(t,\lambda) \leq c(\nu) + m\ \mu_n(t-\lambda)\ I(\lambda \leq t),$$

$$\mu_{n+1}(t) \leq c(\nu) + m \int_0^t \mu_n(t-\lambda)\ dG(\lambda) \leq c(\nu) + mG(t)\ \mu_n(t)\,.$$

If t is so small that $mG(t) < 1$, it therefore follows by iteration that $\mu(t) = \lim \mu_n(t) < \infty$. But if $\mu(t) < \infty$, then 4.1 applied to the inequality

$$Y_{2t} \leq Y_t + \sum_{i=1}^{|Z_t|} Y_t^{t,i}$$

shows easily that $\mu(2t) < \infty$ and therefore $\mu(s) < \infty\ \forall\ s$. □

The following lemma is rather standard and easily proven for example upon integration by parts:

<u>5.6 Lemma</u> (5.5) <u>implies that for any</u> $x, y \geq 0$

$$(5.8)\quad E^x\,Y\,I(Y > y) \leq \int_y^\infty x\ dH(x)$$

$$(5.9)\quad E^x\ Y^2\ I(Y \leq y) \leq \int_0^y x^2\ dH(x) + y(1 - H(y)).$$

5.7 Lemma *Define for some functional* $Y > 0$ *of the process and some fixed* $\delta > 0$

$$S_n = e^{-\alpha n\delta} \sum_{i=1}^{|Z_{n\delta}|} Y^{n\delta,i}, \quad \tilde{S}_n = e^{-\alpha n\delta} \sum_{i=1}^{|Z_{n\delta}|} Y^{n\delta,i}\, I(Y^{n\delta,i} \leqq e^{\alpha n\delta}),$$

$$\eta(x) = E^x Y, \quad \varepsilon_n(x) = E^x(Y\, I(Y > e^{\alpha n\delta})), \quad T_n = E(S_n \mid F_{n\delta}) = e^{-\alpha n\delta} Z_{n\delta}[\eta],$$

$$R_n = e^{-\alpha n\delta} Z_{n\delta}[\varepsilon_n] = E(S_n - \tilde{S}_n \mid F_{n\delta}) = T_n - E(\tilde{S}_n \mid F_{n\delta})$$

Then (5.5) *and* $\int_0^\infty x\, dH(x) < \infty$ *implies that*

$$(5.10) \quad \sum_{n=0}^{\infty} P(S_n \neq \tilde{S}_n) < \infty, \quad \sum_{n=0}^{\infty} \mathrm{Var}\{\tilde{S}_n - T_n + R_n\} < \infty .$$

If furthermore $\int_0^\infty x \log x\, dH(x) < \infty$, *then also* $\sum_{n=0}^\infty E\, R_n < \infty$

Proof One just has to insert (5.5), (5.8), (5.9) in (2.1), (2.2), (2.3). For example,

$$\sum_{n=0}^{\infty} P(S_n \neq \tilde{S}_n) \leqq \sum_{n=0}^{\infty} E\Big(\sum_{i=1}^{|Z_{n\delta}|} P(Y^{n\delta,i} > e^{\alpha n\delta} \mid F_{n\delta})\Big) \leqq$$

$$\sum_{n=0}^{\infty} E\, |Z_{n\delta}|(1 - H(e^{\alpha n\delta})) = \sum_{n=0}^{\infty} O(e^{\alpha n\delta}) \int_{e^{\alpha n\delta}}^{\infty} dH(x) =$$

$$\int_0^\infty O(x)\, dH(x). \ \square$$

Proof of the suffiency of $(x \log x)$. We study the $\{W_t\}_{t \geqq 0}$ - martingale along the discrete subsequence $\{W_n\}_{n=0,1,2,...}$. Let $Y = W_1 = e^{-\alpha} Z_1[V]$ and $\delta = 1$ in 5.7. Then $\eta(x) = V(x)$, $S_n = W_{n+1}$, $T_n = W_n$. Writing $\tilde{W}_{n+1} = \tilde{S}_n$, 5.4, 5.7 implies

$$\sum_{n=0}^{\infty} \mathrm{Var}\{\tilde{W}_{n+1} - W_n + R_n\} < \infty, \quad \sum_{n=0}^{\infty} ER_n < \infty$$

and thus the L^1 - convergence of $\Sigma\{\tilde{W}_{n+1} - W_n + R_n\}$, ΣR_n, $\Sigma\{\tilde{W}_{n+1} - W_n\}$. From this $EW = EW_0$ follows exactly as in § 2. □

Before discussing the problem of the necessity of (x log x), we give the proof of (5.4).

5.8 Lemma Let $M = \sup_{t \geq 0} e^{-\alpha t} |Z_t|$. Then $M < \infty$.

Proof (H. Kesten, private communication. See also [3]). Since W exists, it is clear that $\widetilde{M} = \sup_{t \geq 0} W_t < \infty$ a.s. If $\inf_{x \geq 0} V(x) = c > 0$, then $|Z_t| \leq c^{-1} Z_t[V]$ and thus $M \leq c^{-1} \widetilde{M} < \infty$. In the general case, we always have $V(x) \geq \gamma > 0$ when $0 \leq x < 1$. Any individual alive at time t, $n \leq t \leq n+1$, was alive and of age at most 1 at one of the times $0,1,\ldots,n,t$. Thus

$$|Z_t| \leq \sum_{k=0}^{n} Z_k[I_{[0,1]}] + Z_t[I_{[0,1]}] \leq$$

$$\gamma^{-1}\left(\sum_{k=0}^{n} Z_k[V] + Z_t[V]\right) \leq \gamma^{-1} \widetilde{M}\left(\sum_{k=0}^{n} e^{\alpha k} + e^{\alpha t}\right) = \widetilde{M}\, O(e^{\alpha t})$$

and the assertion follows. □

5.9 Lemma In the notation of 5.7, $\int_0^\infty x\, dH(x) < \infty$ implies that $S_n - T_n \to 0$.

Proof (5.10) implies that $S_n = \widetilde{S}_n$ for n large and that $\widetilde{S}_n - T_n + R_n \to 0$ so we only have to prove $R_n \to 0$. But from (5.8)

$$0 \leq R_n \leq M \int_{e^{\alpha n \delta}}^{\infty} y\, dH(y) \to 0. \;\square$$

5.10 Lemma If $\eta \in B$ satisfies (5.2), then for any $\delta > 0$ a.s.

(5.11) $e^{-\alpha n \delta} Z_{n\delta}[\eta] \to W A[\eta]$.

Proof Let

$$Y_m = e^{-\alpha m \delta} Z_{m\delta}[\eta],\; \eta_m(x) = E^x Y_m,\; c_m = \sup_{0 \leq x < \infty} |\eta_m(x) - V(x) A[\eta]| .$$

In the notation of 5.7 we get, using 5.9,

$$\overline{\lim}\, e^{-\alpha n \delta} Z_{n\delta}[\eta] = \overline{\lim}\, S_n = \overline{\lim}\, T_n \leq$$

$$\overline{\lim}\ e^{-\alpha n\delta}\ \{Z_{n\delta}[V]\ A\ [\eta] + Z_{n\delta}[1]\quad c_m\} \leqq W A [\eta] + M c_m\ .$$

As $m \to \infty$, $\overline{\lim} \leqq$ in (5.10) follows. $\overline{\lim} \geqq$ is similar. □

<u>Proof</u> of (5.4). We first remark, that (5.11) holds whenever $\eta \in B$ is continuous a.e. This follows since we have weak convergence of the (random) measure $e^{-\alpha t}\ Z_t$ to WA for all realizations of the process such that (5.11) holds for $\eta(x) = I(0 \leqq x < a)$, $a = \infty$ or a rational, cf. [8]. The same argument shows that it suffices to establish (5.4) for an η of this specific form. Let then for $\delta > 0$

$$\underline{Y}_\delta = \inf_{0 \leqq t \leqq \delta} Z_t[\eta],\ \overline{Y}_\delta = \sup_{0 \leqq t \leqq \delta} Z_t[\eta],\ \underline{\xi}_\delta(x) = E^x \underline{Y}_\delta,\ \overline{\xi}_\delta(x) = E^x \overline{Y}_\delta .$$

Obviously

$$\underline{\xi}_\delta(x) \geqq I(x + \delta < a)\ P^x(\lambda > \delta) = \underline{\xi}^*_\delta(x)$$

$$\overline{\xi}_\delta(x) \leqq I(x < a)\ P^x(\lambda > \delta) + c\ P^x(\lambda \leqq t) = \overline{\xi}^*_\delta(x)$$

(say), using 5.2, 5.4 for the last estimate. Since $\underline{\xi}^*_\delta$, $\overline{\xi}^*_\delta$ are a.e. continuous, we get from 5.9, 5.10

$$\overline{\lim_{t\to\infty}}\ e^{-\alpha t}\ Z_t[\eta] \leqq \overline{\lim_{n\to\infty}}\ e^{-\alpha n\delta} \sum_{i=1}^{|Z_{n\delta}|} \overline{Y}^{n\delta,i} =$$

$$\overline{\lim_{t\to\infty}}\ e^{-\alpha n\delta}\ Z_{n\delta}[\overline{\xi}_\delta] \leqq \overline{\lim_{n\to\infty}}\ e^{-\alpha n\delta}\ Z_{n\delta}[\overline{\xi}^*_\delta] = W A [\overline{\xi}^*_\delta]\ .$$

If we take the paths right-continuous, then $P^x(\lambda \leqq \delta) \to 0$, $\delta \downarrow 0$, so that $\overline{\xi}^*_\delta \to \eta$. This proves $\overline{\lim} \leqq$ in (5.4) and $\underline{\lim} \geqq$ is similar. □

It remains to prove that $W = 0$ if $(x \log x)$ fails. A short and self-contained proof of this fact is given in [5]. We present here a different proof along the lines of § 2.

<u>5.11 Lemma</u> <u>There exists a set</u> $B \subseteqq [0,\infty)$ <u>of positive</u> A-<u>measure and</u> $c_1 > 0$, $c_2 < \infty$ <u>such that for all</u> $x \in B$, $u > 0$

$$E^x W_1\ I(W_1 > u) \geqq c_1 \int_{c_2 u}^{\infty} x\ dF(x)\ .$$

Proof Let $\gamma_1, \gamma_2, \ldots$ denote constants with $0 < \gamma_i < \infty$. We choose B such that

$$(5.12) \quad \inf_{x \in B} P^x(\lambda \leqq 1) = \inf_{x \in B} \frac{G(x+1 - G(x)}{1 - G(x)} = \gamma_1 > 0 .$$

For example, choose first $y > 0$ in the support of G and next $z > 0$ such that $y - 1 < z < y$ and that both z and z+1 are continuity points of G. Then for x in a suitable open neighbourhood B of z (which has positive A-measure)

$$\frac{G(x+1) - G(x)}{1 - G(x)} \geqq \frac{1}{2} \frac{G(z+1) - G(z)}{1 - G(z)} > 0 .$$

We next remark, that if $S_N = U_1 + \ldots + U_N$ with the U_i i.i.d., $U_i \geqq 0$, $EU_i > 0$, then for some γ_2, γ_3 and all N,u

$$(5.13) \quad E\, S_N I(S_N > u) \geqq \gamma_2 N I(N > \gamma_3 u) .$$

To see this, choose γ_4, γ_5 such that $P(S_{N-1}/N > \gamma_4) \geqq \gamma_5$ for all $N \geqq 2$ and note that a lower bound for the left-hand side of (5.13) is

$$N EU_1 P(U_2 + \ldots + U_N > u) \geqq N EU_1 I(N > u/\gamma_4) \gamma_5 .$$

Now let $N = |Z_\lambda|$, $U_i = \inf_{0 \leqq t \leqq 1} W_t^{\lambda,i}$. Then $W_1 \geqq \gamma_6 S_N$ on $\{\lambda \leqq 1\}$ and thus for $x \in B$

$$E^x W_1 I(W_1 > u) \geqq E^x W_1 I(W_1 > u) I(\lambda \leqq 1) \geqq$$

$$\gamma_6 E^x[I(\lambda \leqq 1) E(S_N I(\gamma_6 S_N > u) \mid N,\lambda)] \geqq$$

$$\gamma_7 E^x[I(\lambda \leqq 1) N I(N > \gamma_8 u)] \geqq \gamma_7 \gamma_1 \int_{\gamma_8 u}^{\infty} x \, dF(x). \ \square$$

Proof of the necessity of (x log x). Choosing $\widetilde{W}_{n+1}$, R_n as for the sufficiency, $\Sigma R_n < \infty$ follows from (5.10) exactly as in § 2. Writing $Z_n = \langle x_1 \ldots x_n \rangle$, we have

$$R_n = e^{-\alpha n} \sum_{i=1}^{|Z_n|} E^{x_i} W_1 I(W_1 > e^{\alpha n}) \geqq e^{-\alpha n} Z_n[I_B] c_1 \int_{c_2 e^{\alpha n}}^{\infty} x \, dF(x).$$

If (x log x) fails then

$$\sum_{n=0}^{\infty} \int_{c_2 e^{\alpha n}}^{\infty} x \, dF(x) = \infty$$

so that on $\{W > 0\}$ it follows from $e^{-\alpha n} Z_n[I_B] \to W A[I_B] > 0$ that $\Sigma R_n = \infty$ a.s. Thus $P(W > 0) = 0$. □

REFERENCES

[1] S. Asmussen, Convergence rates for branching processes, Ann. Probability, 4, 139-146 (1976).

[2] S. Asmussen & H. Hering, Strong limit theorems for general supercritical branching process with applications to branching diffusions, Z. Wahrscheinlichkeitstheorie verw. Geb., 36, 195-212 (1976).

[3] S. Asmussen & H. Hering, Strong limit theorems for supercritical immigration-branching processes, Math. Scand., 39, 327-342 (1976).

[4] K.B. Athreya, On the equivalence of conditions on a branching process in continuous time and on its offspring distribution, J. Math. Kyoto Univ., 9, 41-53 (1969).

[5] K.B. Athreya, On the supercritical one-dimensional age dependent branching processes, Ann. Math. Statist., 40, 743-763 (1969).

[6] K.B. Athreya, A note on a functional equation arising in Galton-Watson branching processes, J. Appl. Probability, 8, 589-598 (1971).

[7] K.B. Athreya & N. Kaplan, Convergence of the age distribution in the one-dimensional supercritical age-dependent branching process, Ann. Probability, 4, 38-50 (1976).

[8] P. Billingsley, Convergence of probability measures, Wiley, New York (1968).

[9] N.H. Bingham & R.A. Doney, Asymptotic properties of supercritical branching processes I: The Galton-Watson process, Advances in Appl. Probability, 6, 711-731 (1974).

[10] T.J.I'a Bromwich, An introduction to the theory of infinite series, Macmillan, London (1908).

[11] W. Feller, An introduction to probability theory and its applications II, 2nd Ed., Wiley, New York (1971).

[12] T.E. Harris, The theory of branching processes, Springer, Berlin (1963).

[13] H. Kesten & B.P. Stigum, A limit theorem for multidimensional Galton-Watson processes, Ann. Math. Statist., 37, 1463-1481 (1966).

[14] J. Neveu, Mathematical foundations of the calculus of probability, Holden-Day, San Fransisco (1965).

ADDITIVE PROPERTY AND ITS APPLICATIONS IN BRANCHING PROCESSES

K. B. Athreya

Department of Applied Mathematics
Indian Institute of Science
Bangalore, India

and

N. Kaplan

Biometry Branch, National Institute
of Environmental Health Sciences
Research Triangle Park, North Carolina, USA

1. INTRODUCTION

One of the basic assumptions of most of the branching process models is that of the independence of lines of descent. Even in the random environments model this assumption holds once we condition on the entire environment sequence (10,11). This independence of lines of descent leads to what may be called the additive property in branching processes. For the simple branching process (Galton-Watson process of a single type in discrete time) it states that for any n, the stochastic process $\{Z_{n+m};\ m=0,1,2,\ldots,\}$ when conditioned on $\{Z_0, Z_1,\ldots,Z_n\}$ is equivalent to

$$\sum_{j=1}^{Z_n} Z_m^{(j)} \ ; \quad m=0,1,2,\ldots,$$

where $\{Z_m^{(j)};\ m=0,1,2,\ldots\}$ for $j=1,2,\ldots$ are independent copies of $\{Z_m;\ m=0,1,2,\ldots\}$ with $Z_o=1$ w.p.1. and independent of Z_n as well. It would be reasonable to expect that one could exploit this in conjunction with classical results from the theory of sums of independent random variables to obtain some of the limit properties of many branching models.

The purpose of this paper is to substantiate this point with many examples. In the next section we formulate the additive property for a general branching model that includes almost all the familiar models. In the third section we apply it to the multitype Galton-Watson process while in section 4 we use it to prove the convergence of the age distribution in the supercritical and critical one dimensional age dependent branching processes. The usefulness of this technique is well illustrated here by the simplicity of the proofs as well as the rather minimal nature of the hypotheses. The final section deals with branching random walks.

2. THE ADDITIVE PROPERTY AND ITS USES

Most of the branching models that have the property of independence of lines of descent are special cases of the following general model. Let $(Y,\mathcal{Y})$ be a measurable space. For $n \geq 1$, let $Y^{(n)}$ be the symmetrized n-fold Cartesian product of Y with itself (i.e., identify two elements of Y^n if they contain the same elements). Let θ be an extra point and set $Y^{(0)}=\{\theta\}$. Let

$\hat{Y} = \bigcup_0^\infty Y^{(n)}$ and $\hat{\mathcal{Y}}$ be the σ-algebra of subsets of $\hat{Y}$ induced by $\mathcal{Y}$. By a discrete time branching process with type space $(Y,\mathcal{Y})$ we mean a Markov chain $\{Z_n: n=0,1,2,...\}$ with stationary transition probabilities and with state space $(\hat{Y},\hat{\mathcal{Y}})$ that satisfies the following branching property: For every decomposition of Y into measurable sets $A_1,A_2,...,A_k$,

$$\begin{aligned} &P\left(I_{A_i}(Z_1) = r_i;\ i=1,2,...,k \mid Z_0 = (y_1 \ldots y_m)\right) \\ &= \prod_{j=1}^{m} P\left(I_{A_i}(Z_1) = r_{ij}\ ;\ i=1,2,...,n \mid Z_0 = (y_j)\right) \end{aligned} \tag{2.1}$$

where the summation on the right is over all (r_{ij}) such that

$$\sum_{j=1}^{m} r_{ij} = r_i \quad \text{for each} \quad i=1,2,...,k,$$

and

$$P\left(I_Y(Z_1) = 0 \mid Z_o = (\theta) = 1\right)$$

where

$$I_A(\hat{y}) = \begin{cases} 0 & \text{if} \quad \hat{y} = (\theta) \\ \sum_{i=1}^{m} I_A(y_i) & \text{if} \quad \hat{y} = (y_1,...,y_m) \end{cases}$$

(For any set A, I_A is its indicator function.) It is clear from the definition of the process that the total data for the problem is the family of probability distributions $\{P(y;\cdot)\}$ where $P(y;E) \equiv P(Z_1 \in E \mid Z_o = (y))$ and E is of the form $\{\hat{y};\ I_{A_i}(\hat{y})=r_i;\ i=1,2,...,k\}$ with $A_1, A_2,...,A_k$ a decomposition of Y into $\mathcal{Y}$ measurable sets and $r_1,...,r_k$ non-negative integers. Thus, $P(y;\cdot)$ denotes the first generation offspring distribution of a particle of type y.

If Y is a singleton then $\hat{Y}$ may be identified with non-negative integers, θ with 0 and $\{Z_n: n=0,1,2,...,\}$ becomes the ordinary Galton-Watson process. Similarly, if Y is a finite set $\{1,2,\ldots,p\}$ we revert to multitype Galton-Watson process (see Chapts. I and II of Athreya-Ney [9]).

By a continuous time branching process with type space $(Y,\mathcal{Y})$ we mean a Markov process $\{Z_t;\ t \geq 0\}$ with stationary transition probabilities and with state space $(\hat{Y},\hat{\mathcal{Y}})$ that satisfies the branching property: For every decomposition of Y into Y-measurable sets, $A_1, A_2,\ldots,A_k$ and every $t \geq 0$,

$$P\left(I_{A_i}(Z_t) = r_i\ ;\ i=1,2,\ldots,k \mid Z_0 = (y_1,\ldots,y_m)\right)$$
$$= \prod_{j=1}^{m} P\left(I_{A_i}(Z_t) = r_{ij}\ ;\ i=1,2,\ldots,k \mid Z_0 = (y_j)\right) \tag{2.2}$$

where the notation is the same as in the discrete time case. Further, $P(I_Y(Z_t) = 0 \mid Z_0 = (\theta)) = 1$ for all $t \geq 0$. When Y is a singleton or a finite set we recover the single or multitype Markov branching process (see Chapts. III and IV of [9]).

For those cases where Y is finite, the distributon of the process is completely specified once the lifetime and offspring distributions are known. For a more general Y, additional information is needed. For example, if particles move during their lifetimes as a Markov process and at death produce offspring according to a distribution, one needs to know the nonbranching Markov process, a lifetime distribution and a branching kernel. We refer the reader to Ikeda et al. [16, 17, 18] for a discussion of these ideas. It suffices to say here that the model defined here is broad enough to include the so called (T_t^0, k,Π) process, branching random walks, branching diffusions, Crump-Mode-Jagers processes, single and multitype age dependent branching processes, etc.

Consider now the discrete time process as defined by (2.1). As in Harris [14] we could take as our underlying sample space, the space of all family histories. It is then possible to construct an increasing sequence $\{F_n\}$ of σ-algebras such that F_n has all the information about family histories up to time n. In particular, the branching property, as defined in (2.1), implies that the stochastic process $\{Z_{n+m}: m=0,1,2,\ldots,\}$ when conditioned on F_n is equivalent to

$$\left\{\sum_{j=1}^{|Z_n|} Z_m^{(j)}\ , \quad m=0,1,2,\ldots\right\}$$

where $\{Z_m^{(j)}\ ; m=0,1,2,\ldots\}$ are independent branching processes. More precisely, for positive integers n and m, there exists random variables $\{Z_{n,m}^{(j)}\ j=1,2,\ldots,|Z_n|\}$ ($|Z_n|$ is the cardinality of Z_n) such that

i) they are F_{n+m} measurable,

ii) when conditioned on F_n are independent,

iii) if $Z_n \neq (\theta)$ then for each bounded measurable f on $(\hat{Y},\hat{y})$ $P(f(Z_{n,m}^{(j)}) \le x|F_n) = P(f(Z_m) \le x \mid Z_0 = y_j)$ where $Z_n = (y_1, y_2,\ldots,y_{|Z_n|})$ and if $Z_n = \theta$, $P(f(Z_{n,m}^{j}) = 0 \mid F_n) = 1$ for all j where

$$f(\hat{y}) = \begin{cases} 0 & \text{if } \hat{y} = (\theta) \\ \sum_{i=1}^{n} f(y_i) & \text{if } \hat{y} = (y_1,\ldots,y_n) \\ & \text{for } 1 \le n < \infty \end{cases}$$

iv) $f(Z_{n+m}) = \sum_{j=1}^{|Z_n|} f(Z_{n,m}^{(j)})$ w.p.1. for any initial distribution Z_0

We shall refer to the above as the <u>additive property</u> of the branching process.

For the continuous time model we need to ensure first that there is no explosion in finite time and next obtain a separable version of the process to meet the necessary measurability requirements. However, assuming non explosion we could certainly obtain the additive property for any skeleton $\{n\delta: n=0,1,2,...\}$ with $0<\delta<\infty$ and for many limit theorems this is sufficient. In Section 4 the special case of the one-dimensional Bellman-Harris process is considered in greater detail.

We shall now describe some ways in which the additive property could be used to prove limit theorems for branching processes. We shall consider here only the discrete case although what we say carries over just as easily to the continuous time case.

To prove a limit result for Z_n as $n \to \infty$ we first look at Z_{n+m}, use the additive property to express it as a random sum of independent random variables such as

$$Z_{n+m} = \sum_{j=1}^{|Z_n|} Z_{n,m}^{(j)} .$$

Next we let n,m go to infinity in an appropriate fashion. There are three ways of doing this. They are:

i) fix m large and let $n \to \infty$.

ii) let $m \to \infty$ with n but such that $m/n \to 0$ and

iii) let $m \to \infty$ first and then $n \to \infty$.

The first is used to prove limit theorems about ratios of functionals on Z_n like convergence of age and type distributions [4], [12], as well as central limit type results [13]. The second is used to prove central limit type results for linear functionals on branching processes when some parameters belong to certain boundary regions [3, 4, 5]. The third type is used to study rates of convergence [2].

3. MULTITYPE GALTON-WATSON PROCESSES

Let $\{Z_n: n\geq 0\}$ be a p-type, positively regular, supercritical Galton-Watson branching process with no extinction (see, [9] for all definitions). Let $M=(m_{ij})$ be the mean matrix, $\rho>1$, be its largest eigenvalue, u and v left and right eigenvectors of M

for the eigenvalue ρ, η a right eigenvector of M with eigenvalue λ. The following results are well known or have been published recently (see, [3], [9], and [22]).

Theorem 3.1

a) $$\sum_{i=1}^{p} \left| \frac{Z_{ni}}{\langle v, Z_n \rangle} - u_i \right| \to 0 \text{ w.p.1} \quad \left(\langle v, Z_n \rangle = \sum_{i=1}^{p} v_i Z_{ni}\right)$$

b) If the process has finite second moments, then

i) $$|\lambda|^2 > \frac{\rho}{2} \Rightarrow \lim_{n\to\infty} \langle \eta, Z_n \rangle \lambda^{-n} = W_n \text{ exists w.p.1}$$

ii) $$|\lambda|^2 < \frac{\rho}{2} \Rightarrow \frac{\langle \eta, Z_n \rangle}{\sqrt{\langle v, Z_n \rangle}} \xrightarrow{d} N(0, \sigma^2) \text{ for some } 0<\sigma^2<\infty$$

iii) $$|\lambda|^2 = \frac{\rho}{2} \Rightarrow \frac{\langle \eta, Z_n \rangle}{\sqrt{\langle v, Z_n \rangle (\log \langle v, Z_n \rangle)}} \xrightarrow{d} N(0, \sigma^2) \text{ for}$$

some $0<\sigma^2<\infty$.

c) If the process has finite second moments, then w.p.1

i) $$|\lambda|^2 > \frac{\rho}{2} \Rightarrow \overline{\lim_n} \frac{\langle \eta, Z \rangle - \lambda^n W_n}{\sqrt{2\sigma^2 \langle v, Z_n \rangle \log n}} = +1$$

$$\underline{\lim_n} \frac{\langle \eta, Z_n \rangle - \lambda^n W_n}{\sqrt{2\sigma^2 \langle v, Z_n \rangle \log n}} = -1$$

for some $0<\sigma^2<\infty$.

ii) $$|\lambda|^2 < \frac{\rho}{2} \Rightarrow \overline{\lim_n} \frac{\langle \eta, Z_n \rangle}{\sqrt{2\sigma^2 \langle v, Z_n \rangle \log n}} = +1$$

$$\underline{\lim_n} \frac{\langle \eta, Z_n \rangle}{\sqrt{2\sigma^2 \langle v, Z_n \rangle \log n}} = -1$$

for some $0<\sigma^2<\infty$.

iii) $$|\lambda|^2 = \frac{\rho}{2} \Rightarrow \overline{\lim_n} \frac{\langle \eta, Z_n \rangle}{\sqrt{2\sigma^2 \langle v, Z_n \rangle n \log n}} = +1$$

$$\lim_{n} \frac{\langle \eta, Z_n\rangle}{\sqrt{2\sigma^2 \langle v, Z_n\rangle\, n \log n}} = -1$$

for some $0<\sigma^2<\infty$.

Part (a) of the above theorem appears in [9] where convergence in probability is established. It is quoted there that a result of T. Kurtz [23] can be used to prove convergence w.p.1 under a jloqj condition. We shall present a proof of the almost sure convergence just with the finiteness of the mean.

Part (b) is due to Kesten and Stigum [22]. Their tools are quite different and involve some nice truncation arguments. The continuous time analog of (b) was established by Athreya [4] where the techniques of the present paper are introduced.

Part (c) is due to Asmussen [3]. Here the proof involves, besides the additive property, a fair amount of technical machinery. We shall now merely outline the proofs of (a) and (b (i) and (ii)) and refer to the original papers for details of these and the rest of the assertions.

Proof (a) By the additive property write

$$Z_{n+m} = \sum_{i=1}^{p} \sum_{j=1}^{Z_{ni}} Z_{n,m}^{(j)} .$$

For any vector ξ

$$\rho^{-m}\langle \xi, Z_{n+m}\rangle = \sum_{i=1}^{p} \sum_{j=1}^{Z_{ni}} (\langle \xi, Z_{n,m}^{(j)}\rangle - (M^m \xi)_i)\rho^{-m}$$

$$+ \langle M^m \xi\rho^{-m} - \langle \xi, u\rangle v, Z_n\rangle$$

$$+ \langle \xi, u\rangle \langle v, Z_n\rangle$$

Thus,

$$\frac{\langle \xi, Z_{n+m}\rangle}{\langle v, Z_{n+m}\rangle} = \frac{A_{n,m} + B_{n,m} + \langle \xi, u\rangle}{A'_{n,m} + B'_{n,m} + \langle v, u\rangle}$$

where

$$A_{n,m} = \sum_{i=1}^{p} \left(\frac{Z_{ni}}{\langle v, Z_n \rangle} - \frac{1}{Z_{ni}} \right) \sum_{j=1}^{Z_{ni}} (\langle \xi, Z_{n,m}^{(j)} \rangle - (M^m \xi)_i) \rho^{-m}$$

$$B_{n,m} = \left\langle (M^m \xi \rho^{-m} - \langle \xi, u \rangle v), \frac{Z_n}{\langle v, Z_n \rangle} \right\rangle$$

and $A'_{n,m}$ and $B'_{n,m}$ being the same as $A_{n,m}$ and $B_{n,m}$ but with ξ replaced by v. Given an $\varepsilon > 0$ choose m_0 large such that $m \geq m_0$

$$\sum_{i=1}^{p} |(M^m \xi \rho^{-m})_i - \langle \xi, u \rangle v_i| < \varepsilon \qquad (3.1)$$

That is possible as guaranteed by the Perron-Frobenius theory [25]. Fix $m = m_0$ and let $n \to \infty$.

The family of random variables $\{(\langle \xi, Z_{m_0} \rangle - (M^{m_0} \xi), Z_0) \rho^{-m_0}\}$ for $Z_0 = e_i$ $i = 1, 2, \ldots, p$ where e_i is the unit vector in the ith direction satisfy the hypothesis of Lemma 4.2 of the next section. Also by a comparison argument (see Lemma 4.5) one can show that $\underline{\lim}_n \frac{Z_{(n+1)i}}{Z_{ni}} > 1$ w.p.1 for each i and hence by Lemma 4.2 $|A_{n,m}|$ and $|A'_{n,m}| \to 0$ w.p.1 as $n \to \infty$. Using (3.1) we have $|B_{n,m_0}|$ and $|B'_{n,m_0}|$ are bounded by constant multiples of ε and so we are done.

Proof (b) Part (i) follows by noting that if $M\eta = \lambda\mu$ and $|\lambda|^2 < \rho$ then $\langle \eta, Z_n \lambda^{-n} \rangle$ is a L_2-bounded martingale. Part (ii): As in (a) use the additive property to write

$$\rho^{-m/2} \langle \eta, Z_{n+m} \rangle = \sum_{i=1}^{p} \sum_{j=1}^{Z_{ni}} (\langle \eta, Z_{n,m}^{(j)} \rangle - (M^m \eta)_i) \rho^{-m/2} + \langle M^m \eta, Z_n \rangle \rho^{-m/2}$$

So

$$\frac{\langle \eta, Z_{n+m} \rangle}{\sqrt{\langle v, Z_{n+m} \rangle}} = (a_{n,m} + b_{n,m}) \sqrt{\frac{\langle v, Z_n \rangle \rho^m}{\langle v, Z_{n+m} \rangle}}$$

where

$$a_{n,m} = \sum_{i=1}^{p} \sqrt{\frac{Z_{ni}}{\sqrt{\langle v, Z_n \rangle}}} \frac{1}{\sqrt{Z_{ni}}} \sum_{j=1}^{Z_{ni}} (\langle \eta, Z_{n,m}^{(j)} \rangle - (M^m \eta)_i) \rho^{-m/2}$$

$$b_{n,m} = \frac{\langle M^m \eta, Z_n \rangle \rho^{-m/2}}{\sqrt{\langle v, Z_n \rangle}}$$

First one shows that if $|\lambda|^2 < \rho$ then ,

$$\overline{\lim_m} \ \overline{\lim_n} \ E\left(| \langle M^m \eta, Z_n \rangle |^2 \right) \rho^{-(m+n)} = 0 .$$

Next choose m large and fix it and let $n \to \infty$. By appealing to the central limit theorem for random sums of random variables one shows that for m fixed.

$$a_{n,m} \xrightarrow{d} N(0, \sigma_m^2) \quad \text{for some} \quad 0 < \sigma_m^2 < \infty.$$

Finally, one shows that $0 < \lim_m \sigma_m^2 = \sigma^2 < \infty$ to finish the proof.

Part (iii) uses the additive property but still needs a fair amount of technical machinery and so we omit it. See [0] for full detail.

4. SINGLE TYPE BELLMAN-HARRIS BRANCHING PROCESSES

The purpose of this section is to apply the techniques described in Section 2 to the one-dimensional Bellman-Harris process. The parameters for this process are a lifetime distribution of G and an offspring distribution $\{p_k\}_{k \geq 0}$. We will always assume:

$$1 \leq m = \sum k p_k < \infty \qquad |Z(0)| = 1$$

$$G \text{ is non-lattice} \qquad G(0+) = 0 . \tag{4.1}$$

The assumptions in (4.1) are all standard. $m \geq 1$ implies the process is not subcritical and G is assumed nonlattice so that standard renewal theory can be applied.

The basic assumptions for the model are the following. First, the initial particle is of age 0. Second, each particle, I, lives a random length of time T_I, having distribution G and upon death produces a random number of offspring N_I, with distribution $\{p_k\}$. The third and last assumption which guarantees (2.2) is that all the $\{T_I, N_I\}_I$ are independent variables.

In the notation of Section 2, we take $Y = R^+$ and interpret the state of the process at time t, $\hat{y} = (y_1, \ldots, y_{|Z(t)|})$, as the ages of the particles alive at time t. The additive property then implies that if at time t, the state of the process is $(y_1, \ldots, y_{|Z(t)|})$, then the future behavior of the process can be looked upon as a sum of $|Z(t)|$ independent processes, each of which is identical with the original Z process except that the original particle is of age y_i instead of age 0, $1 \leq i \leq |Z(t)|$. An explicit construction of the process in terms of the family trees can be found in Harris [14]. We note in passing that $\{|Z(t)|\}_{t \geq 0}$ is what is usually referred to as the Bellman-Harris process.

Our goal in this section is to study the age distribution of the process. Toward this end we define for $z \geq 0$,

$$Z(z,t) = \sum_{i=1}^{|Z(t)|} I_{[o,z]}(y_i) \quad \text{if} \quad Z(t) = (y_1, \ldots, y_{|Z(t)|})$$

and

$$A(z,t) = \begin{cases} \dfrac{Z(z,t)}{|Z(t)|} & \text{if} \quad |Z(t)| > 0 \\ 0 & \text{otherwise} \end{cases}$$

In words $A(z,t)$ denotes the fraction of the population alive at time t of age $\leq z$.

Our results show that $A(z,t)$ converges to a deterministic distribution $A(z)$. Before stating the theorems, we need to introduce some notation.

Let α denote the solution of

$$m\int_{0}^{\infty} e^{-\alpha u}\, G(du) = 1 . \tag{4.2}$$

If $m\geq 1$, then (4.2) has a unique non-negative solution. α is usually referred to as the Malthusian parameter of the population. Let

$$A(x) = \frac{\int_0^x e^{-\alpha u} 1-G(u)du}{\int_0^\infty e^{-\alpha u} 1-G(u)du} \qquad x \geq 0$$

$$G(x) = \frac{G(x+y) - G(y)}{1 - G(y)} \qquad x,y \geq 0$$

$$V(x) = \int_0^\infty e^{-\alpha u}\, G_x(du)$$

$$n_1 = \frac{\int_0^\infty e^{-\alpha u}(1-G(u))du}{\int_0^\infty ue^{-\alpha u}(1-G(u))du}$$

<u>Theorem</u> <u>4.1</u> <u>Let $m>1$</u>. <u>Assume either</u>

$$\sum_{j=2}^{\infty} j\log j\; p_j < \infty \tag{4.3}$$

<u>or</u>

$$\inf_x G_x(y_0) > 0 \quad \text{for some } y_0 > 0 \tag{4.4}$$

Then

$$\lim_{t\to\infty} \sup_z |A(z,t) - A(z)| = 0 \qquad \text{a.s. } E$$

where

$$E = \left\{ \lim_{t\to\infty} Z(t) = \infty \right\}$$

In the critical case (m=1) it is necessary, as usual, to condition on nonextinction.

Theorem 4.2 *Let m=1. Assume also that*

$$\lim_{t\to\infty} \sup_x [1-G_x(t)] = 0 \tag{4.5}$$

Then for any $\varepsilon>0$,

$$\lim_{t\to\infty} P(\sup_z |A(z,t) - A(z)| > \varepsilon \mid |Z(t)| > 0) = 0.$$

Theorem 4.1 is a sharpened version of Theorem B in [12]. Theorem 4.2 is new. The conditions (4.4) and (4.5) hold for example if G is negative exponential or has bounded support.

Using the continuity of A, the fact that $A(\infty)=1$, and the monotonicity of $A(.,t)$ it is not hard to show that for every $\varepsilon>0$ there exists a $\delta>0$ and integer N such that

$$\sup_z |A(z,t) - A(z)| \le \varepsilon + \sup_{1\le j\le N} |A(j\delta,t) - A(j\delta)|$$

Thus we need only establish the asymptotic behavior of $A(z,t)$ for fixed z. So unless otherwise stated we fix $0<z<\infty$.

To prove Theorems 4.1 and 4.2 it is necessary to consider processes $\{Z(t)\}$ which do not start with a particle of age 0 but rather with one of age y, $0<y<\infty$. To denote this we will use a subscript. Thus, for example, $\{Z_y(t)\}_{t\ge0}$ denotes a B.H. process starting with a particle of age y.

The first step in the proof of the theorems is to study the asymptotic behavior of $M_y(z,t) = E(Z_y(z,t))$. Conditioning on the time of death of the initial particle; one obtains the following integral renewal equation:

$$M_y(z,t) = [1-G_y(t)]\, I_{\{t+y\le z\}} + m \int_0^t M(z,t-u)G_y(du) \quad (4.6)$$

Using (4.6) and the renewal theorem [9] it is not difficult to show

Lemma 4.1:

$$\lim_{t\to\infty} e^{-\alpha t} M_y(z,t) = n_1 V(y) A(z) \quad (4.7)$$

As an immediate corrollary we have

Corollary 4.1. If either $m>1$ or $m=1$ and (4.5) holds, the convergence in (4.7) is uniform in y.

Proof. If $m>1$, then Corollary 1 is just Lemma 1 in [12]. If $m=1$, the result follows from (4.5), Lemma 4.1 and the following inequality:

$$|e^{-\alpha t}M_y(z,t) - n_1 A(z)| \le [1-G_y(t)]e^{-\alpha t} + \int_0^t |e^{-\alpha(t-u)}M(z,t-u) - n_1 A(z)| e^{-\alpha u} G_y(du)$$

Without loss of generality, we will assume from now on that $|Z(t)|>0$, $t>0$. The additive property of the process implies that for any t, $s>0$,

$$Z(z,t+s) = \sum_{i=1}^{|Z(t)|} Z_{y_i}(z,s)$$

where $Z(t) = (y_1,\ldots,y_{|Z(t)|})$ and conditional on $Z(t)$, the $\{Z_{y_i}(z,s)\}$ are stochastically independent.

Proceeding as in Section 2, we write

$$e^{-\alpha s}Z(z,\ t+s) = B(z,t) + C(z,t) + e^{-\alpha s}D(t)n_1A(z) \tag{4.8}$$

where

$$B(z,t) = \frac{1}{|Z(t)|}\sum_{i=1}^{|Z(t)|}[e^{-\alpha s}Z_{y_i}(z,s) - \tilde{M}_{y_i}(z,s)]$$

$$C(z,t) = \frac{1}{|Z(t)|}\sum_{i=1}^{|Z(t)|}[\tilde{M}_{y_i}(z,s) - e^{-\alpha s}n_1V(y_i)A(z)]$$

$$D(t) = \frac{1}{|Z(t)|}\sum_{i=1}^{|Z(t)|}V(y_i)$$

$$(\tilde{M}_y(z,s) = e^{-\alpha s}\ M_y(z,s))$$

Thus,

$$A(z,t+s) = \frac{B(z,t) + C(z,t) + e^{-\alpha s}n_1D(t)A(z)}{B(\infty,t) + C(\infty,t) + e^{-\alpha s}n_1D(t)A(z)} \tag{4.9}$$

In view of Corollary 4.1, we can make $|C(z,t) + C(\infty,t)|$ as small as we want on E, the set of nonextinction, by choosing s sufficiently large. Specify an $\varepsilon>0$ and choose and fix s so that

$$\sup_t |C(z,t) + C(\infty,t)| < \varepsilon/2 \tag{4.10}$$

Note that (4.10) holds for all sample paths for which $|Z(t)| > 0$.

The $D(t)$ term can also be dispensed with quickly. If $m=1$, $D(t)\equiv 1$ and so there is no problem. If $m>1$ and (4.2) holds, then

$$\inf_x V(x) \geq e^{-\alpha y_0} \inf_x G_x(y_0) > 0$$

Thus,

$$\inf_t D(t) > 0 \quad \text{a.s.} \quad E \tag{4.11}$$

and so D(t) can be divided out in (4.9). If only $\sum_{j\geq 2} j\log j\, p_j<\infty$, a more involved argument is needed. Let $\hat{V}(t) = \sum_{i=1}^{|Z(t)|} V(y_i)$. It is well known [9] that $\{e^{-\alpha t}\hat{V}(t)\}_{t\geq 0}$ is a nonnegative martingale and converges to a nondegenerate random variable a.s. E, if (4.2) holds. Thus to prove (4.11) we need only show

$$\limsup_{t\to\infty} e^{-\alpha t}|Z(t)| < \infty \quad \text{a.s. } E \tag{4.12}$$

W.L.O.G. assume $\inf_{0\leq x\leq 1} V(x) = a > 0$. Let $|Z^j(t)|$ denote the number of particles alive at time t which were born in the interval [j, j+1). Then $|Z^j(t)| \leq \frac{1}{a}\hat{V}(j)$ and so

$$e^{-\alpha t}|Z(t)| \leq \frac{1}{a}\sum_{j=1}^{[t]} e^{-\alpha j}\,\hat{V}(j)\, e^{-\alpha(t-j)} + e^{-\alpha t}\hat{V}(t) \tag{4.13}$$

(4.12) follows from (4.13) and the fact that $\sup e^{-\alpha t}\,\hat{V}(t) < \infty$ a.s. The previous argument is given in [1] and is attributed to Kesten. So we see that the proofs of Theorems 1 and 2 reduce to showing $|B(z,t) + B(\infty,t)|$ is asymptotically small in the appropriate sense.

We first consider when m>1. Our attention will center on the case when (4.4) holds since the proof of Theorem 1 when (4.3) is valid can be found in [12]. Let $\delta>0$. In view of the arguments given in [12] it suffices to show

$$\lim_{t\to\infty} |B(z,t) + B(\infty,t)| = 0 \quad \text{a.s.} \tag{4.14}$$

where s and t are of the form $k\delta$ for some integer k. We carry out the argument for B(z,t) only. The argument for $B(\infty,t)$ is the same.

We need the following lemma of some independent interest. See also [21].

Lemma 4.2. Let $\{X_{n1}, X_{n2},\ldots,X_{nk_n}\}$, $n=1,2,\ldots$ be an array of random variables such that

(i) for each n $\{X_{ni}\}_{i=1}^{k_n}$ are independent

(ii) $E(X_{ni})=0$ $1\le i\le k_n$, $n\ge 1$

(iii) $\sup P(|X_{ni}| > x) \le C[1-Q(x)]$ for all large x where C is an absolute constant and Q is a cumulative distribution function

Assume further that $\liminf_{n\to\infty} \frac{k_{n+1}}{k_n} > 1$ and $\int_0^\infty xQ(dx) < \infty$. Then for every $\mu>0$,

$$\sum_{n=1}^{\infty} P\left(|\frac{1}{k_n} \sum_{i=1}^{k_n} X_{ni}| > \mu \right) < \infty \qquad (4.15)$$

Proof. By Lemma 1 of [21] for every $\mu>0$,

$$P\left(|\frac{1}{k_n} \sum_{i=1}^{k_n} X_{ni}| > \mu \right) \le C\left[k_n \int_{k_n}^{\infty} Q(dx) + k_n^{-1} \int_0^{k_n} x^2 Q(dx) \right] \qquad (4.16)$$

where C is a constant depending only on Q and μ. Thus,

$$\sum_{n=1}^{\infty} P\left(|\frac{1}{k_n} \sum_{i=1}^{k_n} X_{ni}| > \mu \right) \le C \sum_{n=1}^{\infty} \left[k_n \int_{k_n}^{\infty} Q(dx) + k_n^{-1} \int_0^{k_n} x^2 Q(dx) \right] \qquad (4.17)$$

$$< C \int_0^\infty \left[\sum_{n\le N(x)} k_n + x^2 \sum_{n\ge N(x)} k_n^{-1} \right] Q(dx)$$

where $N(x) = \sup\{n: k_n \le x\}$. Now write

$$\sum_{n<N(x)} k_n = \sum_{n\le N(x)} \left(\frac{k_n}{k_{N(x)}} \; \frac{k_{N(x)}}{x} \, x \right),$$

and

$$x^2 \sum_{n > N(x)} k_n^{-1} = x\, k_{N(x)+1}^{-1} \left(\sum_{n > N(x)} \frac{k_{N(x)+1}}{k_n} \right) x$$

Since $\liminf \frac{k_{n+1}}{k_n} > 1$, both $\sum_{n \le N(x)} \frac{k_n}{k_{N(x)}}$ and $\sum_{n > N(x)} \frac{k_{N(x)+1}}{k_n}$ are bounded in x. Also $k_{N(x)} \le x < k_{N(x)+1}$.

Thus the R.H.S. of (4.17) converges since $\int_0^\infty xQ(dx) < \infty$.

Lemma 4.3. Let the $\{X_{ni}\}$ be as in Lemma 4.2 and assume (i), (ii), (iii) of Lemma 4.2 hold. Assume further that there exist constants C_1, C_2, ρ_1, ρ_2 such that $0 < C_1, C_2 < \infty$, $1 < \rho_1 < \rho_2 < \infty$ and $C_1 \rho_1^n \le k_n \le C_2 \rho_2^n$ for all large n. Let $p = \operatorname{Log} \rho_2 / \operatorname{Log} \rho_1$ and suppose $\int_0^\infty x^p\, Q(dx) < \infty$. Then (4.15) holds.

Proof. Again we need to show the R.H.S. of (4.16) converges. Note for large x

$$N(x) \begin{cases} \le \log(x/C_1)/\log \rho_1 \\ \ge \log(x/C_2)/\log \rho_2 \end{cases}$$

Thus

$$\sum_{n \le N(x)} k_n \le \text{const. } \rho_2^{N(x)} \le \text{const. } x^p$$

and

$$\sum_{n \ge N(x)} k_n^{-1} \le \text{const. } \rho_1^{-N(x)} \le \text{const. } x^{-1/p}$$

Thus the R.H.S. of (4.15) is bounded by

$$\text{const.} \int_0^\infty [x^p + x^{2-1/p}]\, Q(dx) < \infty$$

since $2-1/p < p$ for all $p>0$.

Lemma 4.4. Assume $\sum_{j\geq 1} j^{\beta} p_j < \infty$; $\beta\geq 1$. Then for each $s>0$ there exists a $Q(.)$ such that $\int_0^{\infty} x^{\beta} Q(dx) < \infty$ and

$$\sup_y P\,(|Z_y(z,s)\, e^{-\alpha s} - M_y(z,s)| \geq x) \leq 1-Q(x),\ x>0.$$

Proof.
$$|Z_y(z,s)e^{-\alpha s} - \tilde{M}_y(z,s)| \leq |Z_y(s)|e^{-\alpha s} + \tilde{M}_y(z,s)$$
$$\leq |Z_y(s)|\, e^{-\alpha s} + D$$

where $D = \sup_y [\tilde{M}_y(z,s)] < \infty$ by Corollary (4.1). So it suffices to find a Q satisfying.

$$\sup_y P(|Z_y(s)| \geq x) \leq 1-Q(x)\ .$$

By conditioning on the time of first death we have

$$P(|Z_y(s)| \geq x) \leq \begin{cases} P\left(\sum_{i=1}^{N} |\overset{i}{Z}(s)| \geq x\right) & P_0=0 \\ \\ P\left(\sum_{i=1}^{N} |\overset{i}{\hat{Z}}(s)| \geq x\right) & P_0\neq 0 \end{cases}$$

where N has distribution $\{p_k\}$,

$\{|\overset{i}{Z}(s)|\}_{i=1},\ldots$ are independent of N and independent copies of $|Z(s)|$.

$\{|\overset{i}{\hat{Z}}(s)|\}_{i=1},\ldots$ are independent of N and independent copies of a Bellman-Harris process with offspring distribution $\left\{\hat{p}_0=0,\ \hat{p}_1=p_0+p_1,\ \hat{p}_j=p_j,\ j\geq 2\right\}$.

Let $Q(x) = P\left(\sum_{i=1}^{N} |\overset{i}{Z}(s)| > x\right)$ if $p_0=0$ or $P\left(\sum_{i=1}^{N} |\overset{i}{\hat{Z}}(s)| > x\right)$ if $p_0\neq 0$. In either case $\int_0^{\infty} x^{\beta} Q(dx) < \infty$. This follows from Jensen's inequality and the fact that $\sum j p_j^{\beta} < \infty$ implies $E[Z(s)^{\beta}] < \infty$ [9].

Lemma 4.5. Assume (4.3). Let $k_0=[y_0/\delta]+1$. Then

$$\liminf_{n\to\infty} \frac{|Z((n+1)k_0\delta)|}{|Z(nk_0\delta)|} > 1. \qquad \text{a.s. } E$$

Proof. Using our decomposition we have

$$|Z((n+1)k_0\delta)| = \sum_{i=1}^{|Z(nk_0\delta)|} Z_{y_i}(k_0\ )|$$

Let $\theta>0$ such that $m-\theta>1$. Let $\{|\hat{Z}(t)|\}$ denote a Bellman-Harris process with the same G but with the offspring distribution truncated at some large N so as to guarantee that the truncated mean $\hat{m}>m-\theta>1$. Then

$$|Z((n+1)k_0\delta)| \geq \sum_{i=1}^{|Z(nk_0\delta)|} [|\hat{Z}_{y_i}(k_0\delta)| - \hat{M}_{y_i}(k_0\delta)]$$

$$+ \sum_{i=1}^{|Z(nk_0\delta)|} \hat{M}_{y_i}(k_0\delta) .$$

By our choice of k_0,

$$\begin{aligned} \inf_x M_x(k_0\delta) &\geq 1-G_x(k_0\delta) + \hat{m} \int_0^{k_0\delta} \hat{M}(k_0\delta-u)G_x(du) \\ &\geq 1 + (\hat{m}-1) \inf_x G_x(y_0) > 1 \qquad (4.18) \\ &\quad + (m-\theta-1) \inf_x (G_x(y_0)) > 1 \end{aligned}$$

Thus we need only show

$$\frac{1}{|Z(nk_0\delta)|} \sum_{i=1}^{|Z(nk_0\delta)|} [|\hat{Z}_{y_i}(k_0\delta)|-\hat{M}_{y_i}(k_0\delta)] \to 0 \text{ a.s.} E \qquad (4.19)$$

(4.19) follows directly from Lemma (4.3) and (4.4) and the extended Borel-Cantelli Lemma providing we can show that the

$\{|Z(nk_0\delta)|\}$ satisfy the conditions of Lemma (4.3) a.s. E. This fortunately is not hard. We have already shown that $\sup_t e^{-\alpha t}|Z(t)| < \infty$ a.s. E. So we need only verify the lower bound. Clearly $|Z(t)| \geq |\hat{Z}(t)|$. Furthermore, $e^{-\alpha t}|\hat{Z}(t)|$ converges a.s. to a nondegenerate random variable on $\hat{E}$, the set of explosion for the $\hat{Z}$ process. Since $\hat{E} \uparrow E$, as the truncation gets smaller and the lower bound in (4.18) is independent of the truncation there is no difficulty in establishing the lower bound necessary to apply Lemma 4.3.

Lemma 4.6. $\lim_{n\to\infty} |B(z, nk_0\delta)| = 0$ a.s. E.

Proof. Apply Lemmas 4.2, 4.4, and 4.5 in conjunction with the extended Borel-Cantelli Lemma.

Lemma 4.7. $\lim_{n\to\infty} |B(z,n\delta)| = 0$ a.s. E .

Proof. Duplicating the arguments that went into the proof of Lemma 4.6 one can show in exactly the same way that

$$\lim_{n\to\infty} |B(z, nk_0\delta + j\delta)| = 0 \qquad \text{a.s. } E \quad j=1,2,\ldots,k_0-1.$$

Lemma 4.7 is then immediate.

This more or less completes the proof of Theorem 1. The remaining details are essentially the same as those in [12].

We now turn to Theorem 2. We need to show for every $\eta>0$,

$$\lim_{t\to\infty} P(|B(z, t)| > \eta \mid |Z(t+s)| > 0) = 0 \qquad (4.20)$$

To prove (4.20) it suffices to show

$$\lim_{t\to\infty} \frac{P(|Z(t)|>0)}{P(|Z(t+s)>0)} < \infty \qquad (4.21)$$

and for every $M>0$

$$\lim_{t\to\infty} P(0 < |Z(t)| \leq M \mid |Z(t)| > 0) = 0 \qquad (4.22)$$

To see that (4.21) and (4.22) are sufficient to prove (4.20) we argue as follows. Observe

$$P(|B(z,t)|>\eta \mid |Z(t+s)|>0) = \frac{P(|B(z,t)|>\eta\ ,\ |Z(t+s)|>0)}{P(Z(t+s)>0)}$$

$$= \frac{P(|B(z,t)|>\eta\ ,\ |Z(t)|>0)}{P(|Z(t)|>0)} \ \frac{P(|Z(t)|>0)}{P(|Z(t+s)|>0)}$$

Because of (4.21) we need only show

$$\lim_{t\to\infty} \frac{P(|B(z,t)|>\eta\ ,\ |Z(t)|>0)}{P(|Z(t)|>0} = 0 .$$

Let $M>0$. Then,

$$P(|B(z,t)|>\eta\ ,\ |Z(t)|>0) < P(|Z(t)|<M) + P(|B(z,t)|>\eta,\ |Z(t)|>M).$$

Conditioning on the age chart at time t and using the inequality in (4.16) we obtain

$$P(|B(z,t)| > \mu,\ |Z(t)| > M) \le CE(I_{|Z(t)|>M}\, P(|B(z,t)| > \eta \mid \{y_i\})$$

$$\le CE(I_{|Z(t)|>M}\, H(\ |Z(t)|\ ))$$

where

$$H(w) = \int_w^\infty xQ(dx) + w^{-1}\int_0^w x^2Q(dx) \qquad w>0$$

Since $\int xQ(dx) < \infty$, $H(w) \to 0$ as $w \to \infty$. So by choosing M sufficiently large we can make $\frac{P(|B(z,t)|>\eta,\ |Z(t)|>0)}{P(|Z(t)|>0)}$ as small as we want. To complete the proof of Theorem 2, we prove (4.21) and (4.22).

Lemma 4.8. Assume $m=1$. Then for each $s>0$

$$\lim_{t\to\infty} P(|Z(t+s)|>0) / P(|Z(t)|>0) = 1.$$

Proof. Since $P(|Z(t+s)| > 0) < P(|Z(t)| > 0)$, we have

$$\limsup_{t\to\infty} \frac{P(|Z(t+s)| > 0)}{P(|Z(t)| > 0)} \le 1$$

We need to prove the reverse inequality. Let $q_x(t) = P(|Z_x(t)|=0)$. Then

$$P(|Z(t+s)| > 0) = 1 - E[\prod_{i=1}^{|Z(s)|} q_{y_i}(t)]$$

where $Z(s) = (y_1,\ldots,y_{|Z(s)|})$. Also note

$$q_x(t) = \int_0^t f(q_0(t-u))G_x(du) \le f(q(t)) \qquad (4.24)$$

$(f(s) = \sum_{j=0}^{\infty} p_j{}^j s,\ |s| \le 1)$. Thus,

$$\frac{P(|Z(t+s)|>0)}{P(|Z(t)|>0)} \ge E\left[\frac{1-(q(t))^{|Z(s)|}}{1-q(t)}\right]$$

$$= \frac{1-f(q(t))}{1-q(t)} E[\sum_{j=0}^{Z(s)-1} f(q(t)^j;\ |Z(s)|>0]$$

Since $q(t) \to 1$ as $t \to \infty$,

$$\frac{1-f(q(t))}{1-q(t)} \to f'(1) = 1$$

and

$$E[\sum_{j=0}^{|Z(s)|-1} f(q(t))^j,\ |Z(s)|>0] \to E[|Z(s)|,\ |Z(s)|>0] = 1.$$

Lemma 4.9. Assume $m=1$ and (4.5). Then for any $M>0$

$$\lim_{t\to\infty} P(|Z(t)| \le M \mid |Z(t)| > 0) = 0 \qquad (4.25)$$

Proof. It follows from (4.24) that for any s,

$$q_x(s) \geq p_0\, G_x(s) \geq p_0 \inf_x G_x(s)$$

Because of (4.4) there exists an s such that $\inf_x G_x(s) > \eta > 0$. Thus

$$P[|Z(t+s)|>0] = E[1 - \prod_{i=1}^{|Z(t)|} q_{y_i}(s)] \leq E[1-(\eta p_0)^{|Z(t)|}] .$$

If for some M (4.25) does not hold, then

$$\limsup_{t\to\infty} \frac{P(|Z(t+s)|>0)}{P(|Z(t)|>0)} \leq [1-(\eta p_0)^M]\, P[|Z(t)|<M \mid |Z(t)|>0]$$
$$+ P[|Z(t)|>M \mid |Z(t)|>0]$$
$$< 1$$

which contradicts Lemma 4.8. So (4.25) must be true for all M. This completes the proof of Theorem 2.

We conclude this section with a statement of the corresponding results for the multitype Bellman-Harris process. Proofs of these as well as the more general case of the multitype Crump-Mode-Jagers process found in Ramamurthy [24].

Let $Z(i,a,t)$ be the number of particles alive at time t that are of type i with age $\leq a$. Let $|Z(t)| = \sum_{i=1}^{p} Z(i,\infty,t)$ be the total number of particles alive at time t.

Let M_{ij} be the mean number of type j offspring produced by one particle of type i. Let $G_i^*(\alpha) = \int_0^\infty e^{-\alpha u} dG_i(u)$ where $G_i(u)$ is the life time distribution of function of a type i particle. Assume there exists $\alpha \geq 0$ such that the pxp matrix $((M_{ij}\, G_i^*(\alpha)))$ has unity as its Perron-Froebenius root. Let u and v be p-vectors with positive entries such that

$$\sum_j m_{ij}\, G_i^*(\alpha)\, v_j = v_i \quad \text{for all } i$$

$$\sum_i m_{ij}\, G_i^*(\alpha)\, u_i = u_j \quad \text{for all } j$$

Let

$$V(i,x) = \sum_k v_k \; m_{ik} \int_0^\infty e^{-\alpha u} \, d\, G_x^{(i)}(u)$$

where

$$G_x^{(i)}(u) = \frac{G_i(x+u)-G_i(x)}{(1-G_i(x))},$$

$$A(i,a) = C\, u_i \int_0^a e^{-\alpha u} \, (1-G_i(u))\, du$$

where C is chosen to make $\sum_{i=1}^{p} A(i,\infty) = 1$. The process is called supercritical if $\alpha>0$ and critical if $\alpha=0$.

Theorem 4.1'. (Supercritical) Assume the process is supercritical. Assume either the 'jlogj' conditions on the offspring distributions or $\inf_x V(i,x) > 0$ for each i. Then, on the set of nonextinction,

$$\frac{Z(i,a,t)}{|Z(t)|} \to A(i,a) \quad \text{as} \quad t \to 0 \text{ w.p.1.}$$

Theorem 4.2'. (Critical) Assume the process is critical. Assume $\sup_{i,x} (1-G_x^{(i)}(t)) \to 0$ as $t \to \infty$. Then, $\forall \varepsilon \geq 0$

$$P\left(\left|\frac{Z(i,a,t)}{|Z(t)|} - A(i,a)\right| > \varepsilon \;\middle|\; |Z(t)| \neq 0\right) \to 1 \quad \text{as} \quad t \to \infty.$$

One final remark. It turns out then in Theorem 4.2 the hypothesis (4.5) can in fact be weakened to (4.2). The proof becomes harder and may be found in [24].

5. BRANCHING RANDOM WALKS

The purpose of this section is to show how the decomposition of

Section 2 can be used to obtain limit results for branching random walks and related models. The simplest of such models is the Galton-Watson random walk. Here we have a supercritical Galton-Watson process with offspring distribution $\{p_k\}_{k\geq 0}$, on which we superimpose the additional structure of random walk on the line. A particle whose parent is at x, moves to x+y and the y's of different particles are i.i.d. with distribution Γ. Let $\mu=\int x\Gamma(dx)$ and $\theta^2=\int x^2\Gamma(dx)-\mu^2$. For any sequence of Borel sets $\{D_n\}$ of R define

$$Z_n(D_n) = \text{number of particles in the nth generation in } D.$$

To simplify notation we put $Z_n(y) = Z_n((-\infty,y])$ and $Z_n=Z_n(\infty)$. The objective is to study the asymptotic behavior of $Z_n(D_n)$ as $n\to\infty$.

Without loss of generality we can assume $\mu=0$, $\theta^2=1$ and $p_0=0$. For $-\infty<y<\infty$, let $y_n=\sqrt{n}\ y$. It was originally conjectured by Harris [14] that $m^{-n}Z_n(y_n)$ converges in probability to $\Phi(y)W$ where

$$\Phi(y) = \int_{-\infty}^{y} \frac{1}{\sqrt{2\pi}}\ e^{-z^2/2}\,dz$$

$$m = \sum_{y=1}^{\infty} jp_j$$

and $W=\lim_{n\to\infty} \frac{Z_n}{m^n}$ which we know to exist. The resonability of this conjecture is fairly evident in view of our decomposition. Since $E(Z_n(y_n)) = m^n G^{*n}(y_n)$, we can write for $s_n<n$

$$m^{-n}Z_n(y_n) = m^{-s_n} \sum_{i=1}^{Z_{s_n}} m^{-(n-s_n)}\ Z^i_{n-s_n}(y_n-z_i)-G^{*(n-s_n)}(y_n-z_i)$$

$$+ m^{-s_n} \sum_{i=1}^{Z_{s_n}} [G^{*(n-s_n)}(y_n - z_i) - \Phi(y)]$$

$$+ m^{-s_n} Z_{s_n} \Phi(y),$$

$$= A_n + B_n + m^{-s_n} Z_{s_n} \Phi(y) .$$

where * denotes convolution, $\{z_i\}$ are the positions of the particles in the s_n generation and $\{Z^i_{n-s_n}(y_n - z_i), i=1,\ldots\}$ conditioned on the $\{z_i\}$ are independent random variables with each $Z^i_{n-s_n}(y_n - z_i)$ having the same distribution as $Z_{n-s_n}(y_n - z_i)$.

The idea now is to pick the $\{s_n\}$ so that $s_n \to \infty$ and $B_n \to 0$. This is accomplished by picking the s_n to be much smaller than n. Note how this differs from our choice in Section 3 where s_n was essentially equal to n. Once having picked the $\{s_n\}$, A_n is shown to be negligible by the usual law of large numbers argument. Using this approach, Kaplan and Assmussen [19] proved the following result.

<u>Theorem 5.1</u>. Let $y_n = \sqrt{n}\, y$ and assume $\sum j(\log j)p_j^{1+\varepsilon} < \infty$ for some $\varepsilon > 0$. Then

$$\lim_{n\to\infty} m^{-n} Z_n(y_n) = \Phi(y)\, W \qquad \text{a.s.} \tag{5.1}$$

It turns out that with very little extra work we can prove Theorem 5.1 assuming only $\sum j \log j p < \infty$. Following the arguments in [19] Lemmas 3 and 4, it's not difficult to show that for any $\infty < y < \infty$

$$\lim_{n\to\infty} m^{-n^5} Z_{n^5}(y_{n^5}) = \Phi(y)\, W \quad \text{a.s.} \tag{5.2}$$

It remains to show that (5.2) implies (5.1). For any n choose n_o so that $n_o^5 < n \le (n_o+1)^5$, and let $n_1 = (n_o - 1)^5$. Note that

$$5(n_o-1)^4 \le n-n_1 \le 10(n_o+1)^4$$

Let $\varepsilon>0$, then

$$m^{-n}Z_n(y_n) = Z_{n_1}m^{-n_1} L_1 + L_2 + L_3 + L_4$$

where

$$L_1 = \frac{1}{Z_{n_1}} \sum_{i=1}^{Z_{n_1}} [m^{-(n-n_1)} Z^i_{n-n_1} (y_n - z_i) - G^{*(n-n_1)}(y_n - z_i)]$$

$$L_2 = m^{-n_1} \sum_{i=1}^{Z_{n_1}} G^{*(n-n_1)}(y_n - z_i)\, I_{\{z_i > \sqrt{n_1}\, y\, (1+\varepsilon)\}}$$

$$L_3 = m^{-n_1} \sum_{i=1}^{Z_{n_1}} G^{*(n-n_1)}(y_n - z_i)\, I_{\{z_i < \sqrt{n_1}\, y\, (1-\varepsilon)\}}$$

$$L_4 = m^{-n_1} \sum_{i=1}^{Z_{n_1}} G^{*(n-n_1)}(y_n - z_i)\, I_{\{\sqrt{n_1}\, y(1-\varepsilon) \le z_i \le \sqrt{n_1}\, y(1+\varepsilon)\}}$$

Now by (5.2)

$$\limsup_{n_1\to\infty} |L_4| \le W\{\Phi((1+\varepsilon)y) - \Phi((1-\varepsilon)y)\} \quad \text{a.s.}$$

Consider next L_3. For all i such that $z_i < y(1-\varepsilon)\sqrt{n_1}$

$$\frac{y_n - z_i}{\sqrt{n-n_1}} \ge \frac{\sqrt{n}\, y - \sqrt{n_1}\,(y(1-\varepsilon))}{\sqrt{n-n_1}} \ge \frac{\sqrt{n_1}}{\sqrt{n-n_1}}\, y(1-\varepsilon)$$

$$\to \infty$$

$$\text{as } n_1 \to \infty.$$

Thus

$$|1 - \inf_i G^{*(n-n_1)}(y_n - z_i)| \to 0 \quad \text{as} \quad n_1 \to \infty$$

and so by (5.2)

$$\limsup_{n_1\to\infty} |L_3 - W\Phi(y(1-\varepsilon))| = 0 \qquad \text{a.s.}$$

In a similar fashion

$$\limsup_{n\to\infty} |L_2| = 0 \qquad \text{a.s.}$$

It remains only to show that L_1 converges to zero a.s. To simplify notation let

$$X(j,k) = Z_{j^5}^{-1} \sum_{i=1}^{Z_{j^5}} \left(m^{-(k-j^5)} Z_{(k-j^5)}(y_k - z_i) - G^{*(k-j^5)}(y_k - z_i) \right).$$

Thus, $L_1 = X(n_o - 1, n)$. Let $\delta > 0$. Clearly, for our choice of n,

$$I_{(|L_1| > \delta)} \le \sum_{j=n_o-1}^{\infty} \sum_{k=(j+1)^5+1}^{(j+2)^5} I_{(|X(j,k)| > \delta)}$$

So if we can show

$$\Lambda = \sum_{j=0}^{\infty} \sum_{k=(j+1)^5+1}^{(j+2)^5} I_{(|X(j,k)| > \delta)} < \infty \qquad \text{a.s.} \tag{5.3}$$

we will be done. To prove (5.3) it suffices to show

$$\Lambda_1 = \sum_{j=0}^{\infty} E\left(\sum_{k=(j+1)^5+1}^{(j+2)^5} I_{(|X(j,k)| > \delta)} \,\middle|\, F_{(j+1)^5} \right) < \infty \qquad \text{a.s.} \tag{5.4}$$

where F_j is the σ-algebra generated by the process up to time j. A proof that (5.4) implies (5.3) can be found in [14]. We now prove (5.4). Let $M = \sup_n (Z_n m^{-n})$ and $Q(x) = P(M+2 \le x)$. It follows from arguments in [21] that $\int x\, Q(dx) < \infty$. Furthermore,

$$P\left(\left| m^{-(k-j)^5} Z^i_{(k-j^5)}(y_k - z_i) - G^{*(k-j^5)}(y_k - z_i) \right| > t \right) \le 1 - Q(t)$$

independent of k, j, y_k and z_i. Thus by Lemma 4.2, there exists some constant C such that

$$\Lambda_1 \leq C \int \sum_{j=0}^{\infty} \left[Z_{j^5}(j+1)^4 I_{(Z_{j^5}<x)} + Z_{j^5}^{-1}(j+1)^4 x^2 I_{(Z_{j^5}\geq x)} \right] Q(dx).$$

Using an integral comparison, L'Hospital's rule and the fact that $\lim_n Z_n m^{-n}$ exists and is nonzero, one can show that the integrand above is $O(x)$ a.s. Thus, $\Lambda_1<\infty$ a.s. and so we have Theorem 2 assuming only $\sum j\log j\ p_j < \infty$.

The previous result dealt with Borel sets that were expanding with n. For bounded sets a different normalization is required.

Theorem 5.2. Suppose that $\int|x|^3\Gamma(dx) < \infty$ and $\sum j(\log j)^{3/2+\varepsilon} p_j<\infty$ for some $\varepsilon>0$. Then for $-\infty<a<b<\infty$,

$$(2\pi n)^{1/2}\, m^{-n}\, Z_n[(a,b]] \xrightarrow{a.s.} (b-a)\, W$$

A proof of the theorem can be found in [21]. Presumably using the previous arguments, the moment condition can be weakened, but we have not looked into this.

We now briefly mention some continuous time models. Suppose we have a population which grows as a supercritical Bellman-Harris process with lifetime distribution function G and offspring distribution $\{p_j\}$. Since $m=\sum jp_j>1$, we can always solve $m\int_0^\infty e^{-\alpha z}G(dz)=1$. To describe the motion of particles we have two mechanisms.

Case 1. A particle whose parent was at x at its time of birth moves until it dies according to a standard Brownian Motion starting at x. The motions of different particles are assumed independent.

Case 2. A particle whose parent was at x at its time of birth moves to $x+y$ where it stays until it dies. The y's of different particles are i.i.d. with distribution functions Γ.

In either case, for any Borel set B, let $Z_t(B)$=number of particles in B at time t.

Case 1 was originally studied by Wantanabe [25] in the Markov case using martingale arguments. A recent proof of the general case using our decomposition argument appeared in [21]. We state the result. For any Borel set B, |B| denotes its lebesgue measure.

Theorem 5.3. (Case 1)

(a) Suppose $\sum_j j^2 p_j < \infty$. Then for any set with $|\partial B| = 0$

$$\lim_{t\to\infty} e^{-\alpha t} Z_t(\sqrt{t}\, B) = \Phi(B)\, W \quad \text{a.s.}$$

(b) for any bounded Borel set B with $|\partial B| = 0$,

$$\lim_{t\to\infty} (2\pi t)^{1/2} e^{-\alpha t} Z_t(B) = |B|\, W .$$

The process described in Case 2 has been studied by many authors. The reader should consult [7] for references. Most of this work has dealt with the asymptotic behavior of the mean function. The simplest and most elegant proof of the basic result appeared in [7]. Before we state the result we introduce some notation. Let

$$\nu = \int x\Gamma(dx), \qquad \sigma^2 = \int x^2\Gamma(dx) - \nu^2$$

$$\mu_\alpha = \int_0^\infty t\, m\, e^{-\alpha t}\, G(dt)$$

$$\sigma_\alpha^2 = \int_0^\infty t\, m\, e^{-\alpha t}\, G(dt) - \mu_\alpha^2$$

$$\sigma^2 = \frac{\sigma_-^2}{\mu_\alpha} + \frac{\sigma_\alpha^2 \nu^2}{\mu_\alpha^3}$$

For any $-\infty < \gamma < \infty$, set

$$x_t(\gamma) = t\gamma\mu_\alpha^{-1} + \gamma\sigma\sqrt{t}$$

Theorem 5.4. Assume $\int x^2\Gamma(dx) < \infty$. Then

$$\lim_{t\to\infty} e^{-\alpha t} E\,(Z_t((-\infty, x_t(\gamma))) = n_1\Phi(\gamma) .$$

Just recently [18] we have succeeded, using the decomposition method, in proving the following expected result.

Theorem 5.5. Assume $\int x^2 (dx) < \infty$ and $\sum j \log j p_j\ \infty$. Then

$$\lim_{t\to\infty} \frac{Z_t((-\infty, x_t(\gamma)))}{Z_t((-\infty, \infty))} = \Phi(\gamma) \quad \text{a.s.}$$

The difficulty in proving the theorem can be described as follows. If one was to look at time t at any branch of the family tree, the position of the particle belonging to the branch follows a random walk. However, the number of steps in that walk by time t is random depending on the generation number of the individual picked at time t.

REFERENCES

0. Asmussen, S.(1978). Almost sure behavior of linear functionals of supercritical branching processes. Trans. Amer. Math. Soc. (to appear).
1. Asmussen, S., Herring, H. (1975). Strong limit theorems for supercritical immigration - branching processes. Preprint.
2. Asmussen, S. (1976). Convergence rates for branching processes. Ann. Prob. 4(1): 139-147.
3. Asmussen, S., Keiding, N. (1978). Martingale central limit theorems and asymptotic estimation theory for multiptype branching processes. Adv. App. Prob. (To appear).
4. Athreya, K.B. (1969). Limit theorems for multitype continuous time Markov branching processes I. The case of an eigenvector linear function. Zeit. fur Whar. 12: 320-332.
5. Athreya, K.B. (1969). Limit theorems for multitype continuous time Markov branching processes II. The case of an arbitrary linear functional. Zeit. fur Whar. 13: 204-214.
6. Athreya, K.B. (1970). A simple proof of Kesten and Stigum on supercritical multitype Galton-Watson branching processes. AMS 41: 195-202.

7. Athreya, K.B., Ney, P. (1971). Limit theorems for the means of branching random walks. Sixth Prague Conference on Information Theory.
8. Athreya, K.B. (1971). Some refinements in the theory of supercritical multitype branching processes. Zeit. fur Whar. 20: 47-57.
9. Athreya, K.B., Ney, P. (1972). Branching Processes. Berlin Springer.
10. Athreya, K.B., Karlin, S. (1972a). On branching processes in random environments I - extinction probability. AMS 42: 1499-1520.
11. Athreya, K.B., Karlin, S. (1972b). On branching processes in random environments II - limit theorems. AMS 42: 1843-1858.
12. Athreya, K.B., Kaplan, N. (1976). Convergence of the age-distribution in the one-dimensional age-dependent branching process. Ann. Prob. 4: 38-50.
13. Freedman, D. (1973). Another note on the Borel-Contelli lemma and the strong law with the Poisson approximation as a by product. Ann. Prob. 1(6): 910-926.
14. Harris, T.E. (1963). The Theory of Branching Processes. Berlin Springer.
15. Ikeda, N., Nagasawa, M., Watanabe, S. (1968). Branching Markov processes I. J. of Math. Kyoto Univ. 8: 233-278.
16. Ikeda, N., Nagasawa, M., Watanabe, S. (1968). Branching Markov processes II. J. of Math. Kyoto Univ. 8: 365-410.
17. Ikeda, N., Nagasawa, M., Watanabe, S. (1969). Branching Markov processes. J. of Math. Kyoto Univ. 9: 95-160.
18. Kaplan, N. (1977). A limit theorem for a branching random walk. (to appear).
19. Kaplan, N. And Asmussen, S. (1976). Branching random walks II. Stochastic Processes and Applications 4: 15-31.
20. Karlin, S. (1969). A First Course in Stochastic Processes. Academic Press, New York.
21. Kesten, H., Stigum, B.P. (1966). A limit theorem for multidimensional Galton-Watson processes. AMS 37: 1211-1223.

22. Kesten, H., Stigum, B.P. (1966). Additional limit theorems for undemcomposable multidimensional Galton-Watson processes. AMS 37: 1463-1481.

23. Kurtz, T. (1972). Inequalities for laws of large numbers. AMS 43: 1874-1883.

24. Ramamurthy, (1978). Ph.D. Thesis. Dept. of Applied Math., Indian Instit. of Science, Bangalore, India.

25. Wantanabe, S. (1967). Branching Brownian motion. Unpublished mimeograph. Stanford University.

BRANCHING DIFFUSIONS AND RANDOM MEASURES*

D.DAWSON AND G.IVANOFF

Carleton University
Ottawa,Ontario

1. INTRODUCTION

The main objective of this paper is to give an introduction to some of the problems which arise in the study of infinite systems of particles in R^d which are subject to branching and random spatial motion. These processes arise in the study of the dispersion and fluctuation of spatially distributed populations. For example,

* This research was supported by the National Research Council of Canada.

S. Sawyer [16] has applied them to model the dispersion of a population of rare mutant genes. The finite particle branching diffusions were studied in depth by N. Ikeda, M. Nagasawa and S. Watanabe [7]. The novelty of the infinite particle case arises in the study of the limiting behaviour of the system. Two types of limiting operations are of interest, one in which we look at the behaviour of the number of particles in very large sets and the other in which we consider the number of particles in a fixed set at large times.

In order to keep the exposition simple we restrict our attention to the case of a binary fission process in which the particles move according to a Brownian motion process. The natural setting for the study of infinite particle systems is that of random fields or point processes and we briefly review the latter. The subject of infinite particle branching diffusion processes is at a fairly early stage of development and some of the results involve somewhat tedious computations. Rather than attempting to give an exhaustive survey of current results, we present a sketch of some of the main problems and ideas, emphasizing the central role played by the nonlinear semigroup known as the Skorokhod semigroup.

One of the interesting phenomena of infinite particle branching diffusion processes is the existence of steady state random measures. Two such cases are given special attention, the subcritical case with immigration and the critical case in R^d, $d \geq 3$. In these two cases the existence of a steady state random measure is established and a spatial central limit theorem is proved.

2. PRELIMINARIES ON RANDOM FIELDS AND MEASURES

Let $B(\underset{\sim}{R}^d)$ denote the Borel sets of R^d, d-dimensional Euclidean space, and let Z^+ denote the non-negative integers. Let $\mathcal{M}$ denote the space of all non-negative measures μ on R^d such that $\mu(A) < \infty$ for every compact set A. Let $C_K(R^d)$ denote the family of continuous functions with compact support. For $\mu \in \mathcal{M}$ and $f \in C_K(R^d)$, let

$$\langle \mu, f \rangle \equiv \int f(x)\mu(dx).$$

A sequence $\mu_n \in \mathcal{M}$ is said to converge vaguely to $\mu \in \mathcal{M}$ if $\langle \mu_n, f\rangle \to \langle \mu, f\rangle$ for every $f \in C_K(R^d)$. When $\mathcal{M}$ is furnished with the topology of vague convergence, it becomes a Polish space.

Let $\mathcal{N}$ denote the subspace of $\mathcal{M}$ of Z^+-valued measures. $\mathcal{N}$ is a vaguely closed subset of $\mathcal{M}$. Let $\underset{\sim}{B}(\mathcal{M})$, $\underset{\sim}{B}(\mathcal{N})$ denote the σ-algebras or Borel subsets of $\mathcal{M}, \mathcal{N}$, respectively. $\underset{\sim}{B}(\mathcal{N})$ is generated by all sets of the form

$$\{N | N(A_i) = k_i,\ i = 1,\ldots,n\}$$

where $A_i \in \underset{\sim}{B}(R^d)$, $k_i \in Z^+$, $i=1,\ldots,n$.

For $N \in \mathcal{N}$, a <u>point</u> of the measure N is an element $x \in R^d$ such that $N(\{x\}) > 0$. Intuitively we can regard each point of the process as specifying the location of a particle. If $N(\{x\}) > 1$, the point x is said to be a <u>multiple point</u>, that is, one occupied by more than one particle. Thus we can view N as describing either a measure on R^d or a collection of particles distributed in some way in R^d. In particular if $\{x_j : j \in J\}$, where J is a countable or finite set, denotes the locations of the particles, then for $A \in B(R^d)$,

$$N(A) = \sum_{x_i \in A} \delta(x_i) \tag{2.1}$$

where $\delta(x_i)$ denotes the unit mass at point x_i. Let Λ denote the set of finite or countable collections $\{x_j : j \in J\}$ of elements of R^d such that $\{x_j : j \in J\} \cap A$ is finite for every compact set A. Then (2.1) defines a natural one to one mapping, Ξ, from Λ onto $\mathcal{N}$ and we frequently denote an element of $\mathcal{N}$ by its representation $\{x_j : j \in J\}$. The measurable subsets of Λ are $\underset{\sim}{B}(\Lambda) \equiv \{B | B = \Xi^{-1}C,\ C \in \underset{\sim}{B}(\mathcal{N})\}$.

A <u>random measure</u> is a measurable mapping, X, from a probability space $(\Omega, \mathcal{A}, P)$ into $(\mathcal{M}, \underset{\sim}{B}(\mathcal{M}))$. It induces a probability measure,

P_X, on $(\mathcal{M}, \underset{\sim}{B}(\mathcal{M}))$. P_X is known as the distribution of the random measure.

A random field is a $\mathcal{N}$-valued random measure. (Note that the term point process is more common. However, we reserve the term "process" for the time evolution context).

A sequence of random measures $\{X_n : n \in Z^+\}$ is said to converge in distribution to X if $\langle X_n, f \rangle$ converges in distribution to $\langle X, f \rangle$ for every $f \in C_K(R^d)$.

The finite dimensional distributions generated by a random field are the distributions $P(N(A_1) = k_1, \ldots, N(A_n) = k_n)$, $A_j \in \underset{\sim}{B}(R^d)$, $k_j \in Z^+$, $j = 1, \ldots, n$. A sequence of random fields converges in distribution if and only if the respective finite dimensional distributions converge weakly.

Let X denote a random measure on R^d. Then the Laplace transform of X is defined by

$$L(\varphi) = E(e^{-\langle X, \varphi \rangle}) \quad \text{for} \quad \varphi \in C_K(R^d), \ \varphi \geq 0.$$

(Note that $L(\varphi)$ is also well defined when φ is an indicator function of a bounded set).

Let N denote a random field on R^d. Then the probability generating functional (P.G.F.) is defined by

$$\begin{aligned} G(f) &= E(\mathrm{Exp}(\langle N, \log f(x) \rangle)) \\ &= E(\prod_{i=1}^{\infty} f(x_i)) \end{aligned} \tag{2.2}$$

where $0 \leq f \leq 1$ and $(1-f) \in C_K(R^d)$.

Note that if $\varphi \in C_K(R^d)$, $\varphi \geq 0$, then $1-e^{-\varphi} \in C_K(R^d)$ and $0 \leq 1-e^{-\varphi} \leq 1$. Moreover

$$G(e^{-\varphi}) = L(\varphi).$$

The kth order factorial moment distribution is defined by $M_k = E(N_{(k)})$ where

$$N_{(k)}(A_1 \times \ldots \times A_k | \{x_i\}) = \sum_{i_1 \neq i_2 \neq \ldots \neq i_k} \chi_{A_1 \times \ldots \times A_k}(x_{i_1}, \ldots, x_{i_k})$$

where $A_1, \ldots, A_k \in B(R^d)$ and $\chi_{A_1 \times \ldots \times A_k}(\cdot)$ is the indicator function of $A_1 \times \ldots \times A_k$.

Proposition 2.1

Let $\xi_i(x) = \chi_{A_i}(x)$, $i=1,\ldots,k$. Then

$$M_{(k)}(A_1\times..\times A_k) = \lim_{\eta\uparrow 1}\left\{\frac{\partial^k}{\partial\lambda_1\ldots\partial\lambda_k} G(\eta + \sum_{i=1}^{k}\lambda_i\xi_i)\right\}_{\lambda_1=\ldots=\lambda_k=0}$$

PROOF

Refer to Moyal [14].

The Poisson random field (PRF) with intensity measure $\lambda(.)$ is a random field whose Laplace transform is given by

$$L(\varphi) = \operatorname{Exp}[\int(e^{-\varphi(x)}-1)\lambda(dx)] \tag{2.3}$$

and its P.G.F. is $G(f) = \operatorname{Exp}(\int(f(x)-1)\lambda(dx))$.

The Poisson random field has independent increments and the number of particles in a set A has a Poisson distribution with parameter $\lambda(A)$.

A random measure X (random field) is infinitely divisible if for every r, $r = 1,2,3,\ldots$ there exists a set of independent identically distributed random measures (random fields) $X_{r,k}$, $k=1,\ldots,r$, such that X has the same distribution as $\Sigma_{k=1}^{r} X_{r,k}$.

Proposition 2.2

A random field is infinitely divisible if and only if its P.G.F. has the form

$$G(\xi) = \operatorname{Exp}[\int_{\eta}[\operatorname{Exp}\langle N,\log\xi\rangle-1]\,\Lambda(dN)] \tag{2.4}$$

where $\Lambda(.)$ is a non-negative measure on η such that $\Lambda(\{\phi\}) = 0$, $\Lambda(\{N:N(A) > 0\}) < \infty$ for all bounded sets $A \in \underset{\sim}{B}(R^d)$. Λ is called the KLM (Kerstan-Lee-Matthes) measure of the random field.

PROOF

Refer to Westcott [18; Theorem 5].

An interesting example of an infinitely divisible random field is a Poisson cluster field. In this model each particle of a Poisson random field with intensity measure $\lambda(.)$ is replaced by a random cluster of particles, that is, a finite random field centered at the location of the original particle. Then the PGF of the cluster field, $G_{CL}(.)$, is given by

$$G_{CL}(f) = \mathrm{Exp}[\int (G_x(f)-1)\lambda(dx)] \tag{2.5}$$

where

$$G_x(f) = \int_{\eta} \mathrm{Exp}\langle N, \log f\rangle P_x(dN)$$

where $P_x(.)$, a probability measure on η, is the distribution of the random cluster generated by a particle located at x. In this case the KLM measure is given by $\Lambda(.) = \int P_x(.)\lambda(dx)$.

Another type of random field which we have occasion to use is the doubly stochastic Poisson random field. This random field is obtained by taking a Poisson random field whose intensity measure is again a random measure. The P.G.F., $G_{DS}(.)$, and the Laplace transform, $L_{DS}(.)$ of the doubly stochastic Poisson random field are given by

$$G_{DS}(f) = L_I(1-f) \tag{2.6}$$

$$L_{DS}(\varphi) = L_I(1-e^{-\varphi}) \tag{2.7}$$

where $L_I(.)$ is the Laplace transform of the intensity random measure.

The reader is referred to the survey papers by Daley and Verre-Jones [3] and Jagers [9] for a more detailed exposition of the theory of point processes and random measures.

3. MULTIPLICATIVE MARKOV PARTICLE SYSTEMS

A Markov measure-valued process is a Markov process $\{X_t : t \geq 0\}$ with state space $(\mathcal{M}, \mathcal{B}(\mathcal{M}))$. A Markov particle system is a Markov process, $\{N_t : t \geq 0\}$, with state space $(\mathcal{N}, \underset{\sim}{B}(\mathcal{N}))$ and time homogeneous transition probabilities; $P_t(.|\{x_j\})$. For each $t \geq 0$, $\{x_j\} \in \Lambda$, $P_t(.|\{x_j\})$ is a probability measure on $(\mathcal{N}, \underset{\sim}{B}(\mathcal{N}))$ and for each $A \in \underset{\sim}{B}(\mathcal{N})$, $P_t(A|.)$ is a $\underset{\sim}{B}(\Lambda)$-measurable function. Thus for each $t \geq 0$, N_t is a random field on R^d.

A Markov particle system is multiplicative if for all $\nu_1, \nu_2 \in \mathcal{N}$, $t \geq 0$,

$$P_t(.|\nu_1+\nu_2) = P_t(.|\nu_1)*P_t(.|\nu_2)$$

where $*$ denotes convolution. In other words the distribution of the process with initial condition $\nu_1+\nu_2$ is equal to the distribution of the sum of independent versions of the process with initial conditions ν_1 and ν_2, respectively. We set

$$G_t(\xi|\{x_j\}) = \int_{\mathcal{N}} \operatorname{Exp}(\langle N, \log\xi\rangle) P_t(dN|\{x_j\}).$$

The Markov process $(N_t : t \geq 0)$ is multiplicative if and only if

$$G_t(\xi|\{x_j : j \in J\}) = \prod_{j \in J} G_t(\xi|x_j)$$

where $G_t(\xi|x)$ is the PGF for one initial particle at $x \in R^d$. The PGF of a multiplicative Markov particle system satisfies the semigroup

$$G_{t+s}(\xi|x) = G_s(G_t(\xi|.)|x), \qquad s,t \geq 0, \tag{3.1}$$

$$G_0(\xi|x) = \xi(x).$$

A multiplicative Markov particle system is uniquely determined by a <u>transition probability generating functional</u> (TPGF) satisfying (3.1).

In the remainder of this section we restrict our attention to <u>finite particle systems</u>, that is, we assume

$$P(N_t(R^d) < \infty) = 1 \quad \text{for } t \geq 0.$$

We also assume that $\{N_t : t \geq 0\}$ is a strong Markov process. The process is characterized by the transition probability $P_t(A|x^k)$ of a transition from $x^k \in (R^d)^k$ to $A \in \underset{\sim}{B}(\Lambda_F)$ where $\Lambda_F = \cup_{n=0}^{\infty} (R^d)^n$. $x \in \Lambda_F$ is of the form $\{x_1, \ldots, x_n\}$ and denotes a finite field of particles located at the points $(x_1,\ldots,x_n)$. Note that the transition $x^k \to A$ can involve a change in the size of the population; it must do so in sudden "jumps" so the theory of discontinuous Markov processes applies (Moyal [13]).

The transition functions $P_t(A|x^k)$ satisfy the following integral equations

$$P_t(A|x^k) = P_t^o(A|x^k) + \sum_{j=0}^{\infty} \int_0^t \int_{(R^d)^j} P_{t-s}(A|y^j)Q^{(j)}(dy^j,ds|x^k) \tag{3.2}$$

where

(i) $P_t^o(A|x^k)$ is the probability of a transition $x^k \to A$ in $(0,t)$ <u>without jumps</u>

(ii) $Q = \sum_{j=0}^{\infty} Q^{(j)}$ is the joint distribution of the first jump time and the consequent state.

In terms of TPGF's, (3.2) becomes

$$G_t(\xi|x^k) = G_t^o(\xi|x^k) + \sum_{j=0}^{\infty} \int_0^t \int_{(R^d)^j} G_{t-u}(\xi|y^j)Q^{(j)}(dy^j,du|x^k) \tag{3.3}$$

where G_t^o is the P.G.F. corresponding to P^o. Equation (3.2) (or (3.3)) is called the Moyal equation (c.f. Ikeda, Nagasawa, Watanabe [7]) and is true for any discontinuous strong Markov process.

When N_t is a multiplicative point process, its P.G.F. satisfies $G_t(\xi|x^k) = \Pi_{i=1}^k G_t(\xi|x_i)$ where $x^k = \{x_1,\ldots,x_k\}$. Therefore it is sufficient to consider the restriction of the Moyal equation to one dimension.

$$G_t(\xi|x) = G_t^o(\xi|x) + \sum_{j=0}^{\infty} \int_0^t \int_{(R^d)^j} \left[\prod_{i=1}^{j} G_{t-s}(\xi|y_i)\right]Q^{(j)}(dy^j,ds|x). \tag{3.4}$$

Ikeda, Nagasawa and Watanabe [7] called this nonlinear integral equation for $G_t(\xi|x)$ the S-equation (Skorokhod equation, c.f.[17]).

In order to investigate the existence and uniqueness of a solution to (3.4), consider the following regularity conditions:

I. $G_t^o(\xi|x)$ is the P.G.F. of an incomplete transition probability

$P_t^o(A|x)$, $A \subset \Lambda_F$ (incomplete in the sense that $P_t^o(\Lambda_F|x) \le 1$).

$G_{t+s}^o(\xi|x) = G_t^o(G_s^o(\xi|.)|x)$; $s,t \ge 0$, $G_o^o(\xi|x) = \xi(x)$.

II. $Q_t(A|x) = Q(A,[0,t)|x) = \Sigma_{j=0}^{\infty} Q^{(j)}(A^{(j)}, [0,t)|x)$ where $A^{(j)} \equiv A \cap (R^d)^j$. $Q_t(A|.)$ is assumed to be $\underset{\sim}{B}(R^d)$-measurable for each $A \in \underset{\sim}{B}(\Lambda_F)$ and $Q_t(.|x)$ is a probability measure on $\underset{\sim}{B}(\Lambda_F)$ for each $x \in R^d$.

III. (i) $P_t^o(\Lambda_F|x) + Q_t(\Lambda_F|x) = 1$ for each $t \ge 0$ and $x \in R^d$.

(ii) For $s,t \ge 0$, $Q_{t+s}(A|x) = Q_t(A|x) + \int_{R^d} Q_t(A|y)P_s^o(dy|x)$

Theorem 3.1

(Moyal [15, Theorem 3.1]). Let P_t^o and Q_t satisfy conditions I,II,III and $0 \le \xi(x) \le 1$. Then,

(a) the recursive relation

$$G_t^{(o)} \equiv G_t^o$$

$$G_t^{(n+1)}(\xi|x) = G_t^o(\xi|x) + \sum_{j=0}^{\infty} \int_0^t \int_{(R^d)^j} \prod_{i=1}^{j} G_{t-u}^{(n)}(\xi|y_i)Q^{(j)}(dy^j, du|x)$$

defines a non-decreasing sequence $\{G^{(n)}\}$ of functionals which converge to a function $G^{\infty} \le 1$. G^{∞} is the smallest non-negative solution of (3.4) and satisfies (3.1).

(b) G_t^{∞} is the P.G.F. of a possibly incomplete transition probability $P_t^{\infty}(.|.)$.

(c) G_t^{∞} is the unique bounded solution of (3.4) if and only if $P_t^{\infty}(\Lambda_F|x) = 1$ for all $x \in R^d$, $t > 0$.

4. THE SIMPLE BRANCHING DIFFUSION PROCESS

The simple branching diffusion process is a multiplicative Markov particle system in which the particles independently undergo a random motion in R^d and branching. The process is described as follows:

(a) Brownian Spatial Motion. Each particle is assumed to perform an independent Brownian motion in R^d with transition probability density function $p(t,x,y)$ until the time that it branches. Note that

$$p(t,x,y) = (2\pi t)^{-d/2}\exp(-|x-y|^2/2t), \quad x,y \quad R^d, \quad t \quad 0.$$

The semigroup of operators on the space of bounded continuous functions on R^d, $C_B(R^d)$, associated with the Brownian motion diffusion process is defined by

$$T_f f(x) \equiv \int_{R^d} p(t,x,y)f(y)\,dy.$$

(b) <u>The Branching Rate</u>. It is assumed that the probability that a given particle branches in the interval $[t,t+\Delta t)$ is $V\Delta t + o(\Delta t)$ so that the time until the first branch is an exponential random variable with mean V^{-1}. V is called the branching rate.

(c) <u>The Binary Branching Mechanism</u>. When a particle "branches" it dies with probability $1-\alpha$ and is replaced at the same location by two descendents with probability α, $0 < \alpha < 1$. The descendents then act as independent particles.

In the terminology of the general multiplicative Markov particle system,

$$G_t^o(\xi|x) = T_t^o\xi(x) = e^{-Vt}\int_{R^d} \xi(y)p(t,x,y)\,dy$$

$$Q^{(j)}(dy^j,du\ x) = \begin{cases} Ve^{-Vu}(1-\alpha)du, & j=0 \\ Ve^{-Vu}\alpha p(u,x,y)\delta_y^2\,du, & j=2 \\ 0, & \text{otherwise} \end{cases}$$

where δ_y^2 denotes a unit mass at the point $(y,y) \in (R^d)^2$, that is, two particles located at the location $y \in R^d$.

It is easy to verify that if there is one initial particle, there will be finitely many particles with probability one for every finite t and hence the S-equation has a unique bounded solution.

The S-equation for the simple branching diffusion process is

$$G_t(\xi|x) = e^{-Vt}\int\xi(y)p(t,x,y)\,dy + (1-\alpha)(1-e^{-Vt}) + V\alpha\int_0^t\int G^2_{t-u}(\xi|y)e^{-Vu}p(u,x,y)\,dy\,du. \tag{4.1}$$

If $H_t(1-\xi|x) \equiv 1-G_t(\xi|x)$, then H_t satisfies the equation

$$\begin{aligned} H_t(1-\xi|x) &= e^{-Vt}\int(1-\xi(y))p(t,x,y)dy \\ &+ 2V\alpha\int_0^t\int H_{t-u}(1-\xi|y)e^{-Vu}p(u,x,y)\,dydu \\ &- V\alpha\int_0^t\int H^2_{t-u}(1-\xi|y)e^{-Vu}p(u,x,y)\,dydu \end{aligned}$$

$$H_o(1-\xi|x) = 1-\xi(x).$$

In other words,

$$H_t(1-\xi|x) = T^o_tH_o(1-\xi|x) + 2V\alpha\int_0^t T^o_uH_{t-u}(1-\xi|x)du -V\alpha\int_0^t T^o_uH^2_{t-u}(1-\xi|x)du. \tag{4.2}$$

Note that the infinitesimal generator of the semigroup T^o_t is Δ-VI where Δ is the d-dimensional Laplacian and I is the identity operator.

We then obtain the alternate formulation of Equation (4.2), first obtained by Skorokhod [17],

$$\frac{\partial}{\partial t}H_t(1-\xi|x) = (\Delta+V(2\alpha-1)I)H_t(1-\xi|x) - V\alpha H^2_t(1-\xi|x), \tag{4.3}$$

$$H_o(1-\xi|x) = 1-\xi(x).$$

We now exploit the S-equation to obtain expressions for the factorial moment distributions.

Proposition 4.1

(Ivanoff [8, Theorem 3.1]). Let $M_{(k)}(A_1\times\ldots\times A_k,t|x)$ denote the factorial moment distributions of the simple branching diffusion process with one initial particle located at x. Then

$$M_{(k)}(A_1\times\ldots\times A_k,t|x) = \int_{A_1}\cdots\int_{A_k}\hat{P}_k(t,x;u_1,\ldots,u_k)du_1\ldots du_k$$

where the $\hat{P}_k$ are obtained from the following recursive relationships.

$$\hat{P}_1(t,x,y) = e^{V(2\alpha-1)t}p(t,x,y) \tag{4.4}$$

$$\hat{P}_k(t,x;y_1,\ldots,y_k) = V\alpha\int_0^t\int e^{V(2\alpha-1)u}p(u,x,y)f_k(t-u,y,y_1,\ldots,y_k)dydu$$

where

$$f_k(t,y,y_1,\ldots,y_k) = \sum_{j=1}^{k-1}\sum_{\rho\in S_k^j}\hat{P}_j(t,y;y_{\rho(1)},\ldots,y_{\rho(j)})$$

$$\cdot\ \hat{P}_{k-j}(t,y;y_{\rho(j+1)},\ldots,y_{\rho(k)})$$

where "$\sum_{\rho\in S_k^j}$" indicates summation over all possible combinations of k elements into two subsets of size j and $k-j$ respectively.

PROOF

Equation (4.3) can be rewritten as

$$H_t(1-\xi|x) = e^{V(2\alpha-1)t}\int p(t,x,y)(1-\xi(y))dy$$

$$-V\alpha\int_0^t\int e^{V(2\alpha-1)u}p(u,x,y)H_{t-u}^2(1-\xi|y)dydu$$

which implies that

$$G_t(\xi|x) = 1 + e^{V(2\alpha-1)t}\int p(t,x,y)(\xi(y)-1)dy$$

$$+ V\alpha\int_0^t\int e^{V(2\alpha-1)u}p(u,x,y)[G_{t-u}^2(\xi|y)-2G_{t-u}(\xi|y)+1]dydu$$

But, by Proposition 2.1,

$$M_{(k)}(A_1\times\ldots\times A_k,t|x) = \lim_{\eta\uparrow 1}\left\{\frac{\partial^k}{\partial\lambda_1\ldots\partial\lambda_k}G_t(\eta+\sum_{i=1}^k\lambda_i\xi_i|x)\right\}_{\lambda_1=\ldots=\lambda_k=0}$$

$$= \lim_{\eta\uparrow 1}\{\delta_{1k}e^{V(2\alpha-1)t}\int p(t,x,y)\xi(y)dy$$

$$+V\alpha\int_0^t\int e^{V(2\alpha-1)u}p(u,x,y)[2G_{t-u}(\eta|y)\frac{\partial^k}{\partial\lambda_1\ldots\partial\lambda_k}G_{t-u}(\eta+\Sigma\lambda_i\xi_i|y)\Big|_{\lambda_1=\ldots=\lambda_k=0}$$

$$+\sum_{j=1}^{k-1}\sum_{\rho\in S_k^j}\frac{\partial^j}{\partial\lambda_{\rho(1)}\cdots\partial\lambda_{\rho(j)}}G_{t-u}(\eta+\Sigma\lambda_i\xi_i|y) \tag{4.5}$$

$$\cdot\frac{\partial^{k-j}}{\partial\lambda_{\rho(j+1)}\cdots\partial\lambda_{\rho(k)}}G_{t-u}(\eta+\Sigma\lambda_i\xi_i|y)\Big|_{\lambda_1=\ldots=\lambda_k=0}$$

$$-2\frac{\partial^k}{\partial\lambda_1\ldots\partial\lambda_k}G_{t-u}(\eta+\Sigma\lambda_i\xi_i|y)\Big|_{\lambda_1=\ldots=\lambda_k=0}]\}dydu$$

where $\sum_{\rho\in S_k^j}$ indicates summation over all possible combinations of k elements into two subsets of size j and $k-j$, respectively. But $\lim_{\eta\uparrow 1}G_{t-u}(\eta|y) = 1$ and

$$\lim_{\eta\uparrow 1}\left\{\frac{\partial^j}{\partial\lambda_{\rho(1)}\cdots\partial\lambda_{\rho(j)}}G_{t-u}(\eta+\Sigma\lambda_i\xi_i|y)\Big|_{\lambda_1=\ldots=\lambda_k=0}\right\}$$

$$= M_{(j)}(A_{\rho(1)}\times\ldots\times A_{\rho(j)},t-u|y).$$

Thus (4.5) becomes

$$M_{(k)}(A_1\times\ldots\times A_k,t|x) = \delta_{1k}e^{V(2\alpha-1)t}\int_{A_1}p(t,x,y)dy \tag{4.6}$$

$$+ V\alpha\int_o^t\int_{R^d} e^{V(2\alpha-1)u}p(u,x,y)[\sum_{j=1}^{k-1}\sum_{\rho\in S_k^j}M_{(j)}(A_{\rho(1)}\times\ldots\times A_{\rho(j)},t-u|y)$$

$$\cdot M_{(k-j)}(A_{\rho(j+1)},\ldots,A_{\rho(k)},t-u|y)]dydu.$$

Thus a recursive relationship has been obtained for the factorial moment distributions. It can easily be verified that there are corresponding factorial moment density functions which satisfy the recursive relationships (4.4) and the proof is complete.

5. INFINITE PARTICLE BRANCHING DIFFUSIONS

In this section two Markov infinite particle systems are introduced.

5.1 The Branching Random Field

The branching random field is a simple branching diffusion process with an initial Poisson random field of particles in R^d. The initial Poisson random field is assumed to have intensity measure $\lambda(.)$, a Borel measure on R^d.

Theorem 5.1.1

Let $G_t(\xi|x)$ denote the TPGF of the simple branching diffusion process and let $G_t(\xi)$ denote the PGF of the branching random field. Assume that $\int_A \int_{R^d} p(t,x,y)\lambda(dx)dy < \infty$ for every compact set A. Then,

$$G_t(\xi) = \exp[-\int H_t(1-\xi|x)\lambda(dx)] \quad (5.1)$$

where

$$H_t(1-\xi|x) = 1 - G_t(\xi|x).$$

PROOF

Let $\{D_k\}$ be an increasing sequence of compact sets such that $D_k \uparrow R^d$. Assume that the initial PRF is confined to D_k. Thus initially there are finitely many particles with probability one. Let $G_t^k(\xi)$ denote the corresponding PGF.

$$\begin{aligned}
G_t^k(\xi) &= E[\exp\int \log(\xi(x))N_t(dx)] \\
&= E[E[\exp\int \log(\xi(x))N_t(dx)|\{x_1,\ldots,x_j\}]] \\
&= E[G_t(\xi|\{x_1,\ldots,x_j\})] \\
&= E[\prod_{i=1}^{j} G_t(\xi|x_i)] \\
&= \sum_{j=0}^{\infty} \frac{e^{-\lambda(D_k)}}{j!} \cdot (\int_{D_k} G_t(\xi|x)\lambda(x))^j \\
&= \exp[\int_{D_k} (G_t(\xi|x)-1)\lambda(dx)] = \exp[-\int_{D_k} H_t(1-\xi|x)\lambda(dx)].
\end{aligned}$$

Clearly, by the monotone convergence theorem,

$$G_t^k(\xi) \to G_t(\xi) = \exp[-\int H_t(1-\xi|x)\lambda(dx)].$$

It remains to show that $G_t(.)$ is the PGF of a random field. But $G_t(\xi)$ can be rewritten as

$$G_t(\xi) = \exp[\int_{R^d} \{\int_{\eta}(\exp < N, \log\xi > -1)P_t(dN|x)\}\lambda(dx)]$$

$$= \exp[\int_{\eta}(\exp < N, \log\xi > -1)\Lambda_t(dN)]$$

where

$$\Lambda_t(A) = \int_{R^d} P_t(A|x)\lambda(dx), \quad A \in \mathcal{B}(\eta), \quad \phi \notin A \tag{5.2}$$

$$\Lambda_t(\{\phi\}) = 0,$$

and where $P_t(.|x)$ is the transition probability function for the simple branching diffusion process. Thus by Proposition 2.2 it remains to show that for compact sets A,

$$\Lambda_t(N(A) \neq 0) < \infty.$$

Suppose that a single particle alive at $t = 0$ has exactly k progeny alive at time t with probability $q_k(t)$. Note that $\sum_{k=0}^{\infty} k \cdot q_k(t) = e^{V(2\alpha-1)t}$. Now, the probability that any one particular offspring of the original ancestor at x at time 0 is in A at time t is $\int_A p(t,x,y)dy$. Hence the probability that at least one of the offspring of the original particle is in A at time t is

$$P_t(N(A) \neq 0|x) \leq \sum_{k=0}^{\infty} kq_k(t)\int_A p(t,x,y)dy$$

$$= e^{V(2\alpha-1)t}\int_A p(t,x,y)dy.$$

Hence

$$\Lambda_t(N(A) \neq 0) \leq e^{V(2\alpha-1)t}\int_A \int_{R^d} p(t,x,y)(dx)dy$$

$$< \infty$$

and the proof is complete.

Corollary

The branching random field at time t is an infinitely divisible random field with KLM measure Λ_t.

Proposition 5.1.2

(Ivanoff [8, Theorem 3.4]).

$$M_{(k)}(A_1\times\ldots\times A_k,t) = \sum_{j=1}^{k} \int_{(R^d)^j} \sum_{\rho_k(i_1,\ldots,i_j)} M_{(i_1)} (A_{\rho(1)}\times\ldots\times A_{\rho(i_1)},t|x_1)\ldots M_{(i_j)} (A_{\rho(k-i_j)}\times\ldots\times A_{\rho(k)},t|x_j)\lambda(dx_1)\ldots\lambda(dx_j) \tag{5.3}$$

where " $\sum_{\rho_k(i_1,\ldots,i_j)}$ " indicates summation over all possible groupings of k distinguishable objects into j nonempty subsets.

$$M_{(k)}(A_1\times\ldots\times A_k,t) = \int_{A_1}\ldots\int_{A_k} \hat{P}_k(t;y_1,\ldots,y_k)dy_1\ldots dy_k \tag{5.4}$$

where

$$\hat{P}_k(t;y_1,\ldots,y_k) = \sum_{j=1}^{k} \int_{(R^d)^j} \sum_{\rho_k(i_1,\ldots,i_j)}$$

$$\cdot\hat{P}_{i_1}(t,x_1,y_{\rho(1)},\ldots,y_{\rho(i_1)})\ldots\hat{P}_{i_j}(t,x_j;y_{\rho(k-i_j)},\ldots,y_{\rho(k)})$$

$$\cdot\lambda(dx_1)\ldots\lambda(dx_j),$$

and $\hat{P}_k(t,x;y_1,\ldots,y_k)$ is the kth order factorial moment density function of the branching diffusion with one particle initially as given in Equation (4.4).

PROOF

The proof follows by a similar computation to that of Proposition 4.1.

5.2 The Simple Branching Diffusion With Immigration

In this prccess, the domain R^d is initially empty, and immigration of new particles into R^d is allowed. Once a particle is in R^d, it obeys the usual branching diffusion laws.

The immigration is assumed to take place with rate r: in other words, the probability that a new particle is created in an element of space Δx in $[t,t+\Delta t]$ is $r\Delta t|\Delta x| + o(\Delta t|\Delta x|)$, where $|\Delta x|$ denotes the Lebesgue measure of Δx.

Theorem 5.2.1

Let $G_t(\xi|x)$ denote the TPGF of the sample branching diffusion process and let $G_t^I(\xi)$ denote the PGF of the branching diffusion with immigration. Then

$$G_t^I(\xi) = \exp[-r\int_0^t \int H_s(1-\xi|x)dx] \qquad (5.5)$$

where $H_s(1-\xi|x) = 1-G_s(\xi|x)$.

PROOF

This theorem is proved in the same way as was Theorem 5.1.1. Let $\{D_k\}$ be an increasing sequence of compact sets such that $D_k \uparrow R^d$. Assume that the immigration is confined to D_k. Thus, at any finite time, there have been finitely many immigrants with probability one. Let G_t^{Ik} denote the corresponding PGF.

$$\begin{aligned} G_t^{Ik}(\xi) &= E[\exp\int \log(\xi(x))N_t^I(dx)] \\ &= E[E\exp\int \log(\xi(x))N_t^I(dx)\,|\,n \text{ immigrants appear in } D_k \text{ in } [0,t) \text{ with locations and times of immigration given by } \{(x_1,s_1),\ldots,(x_n,s_n)\}] \\ &= E[G_t(\xi|\{(x_1,s_1),\ldots,(x_n,s_n)\}] \\ &= E[\prod_1^n G_{t-s_i}(\xi|x_i)] \end{aligned}$$

$$= \sum_{n=0}^{\infty} \frac{e^{-rt|D_k|}(rt|D_k|)^n}{n!} \prod_{i=1}^{n} \left(\int_o^t \int_{D_k} G_{t-s_i}(\xi|x_i)\frac{dx_i}{|D_k|}\frac{ds_i}{t}\right)$$

$$= e^{-rt|D_k|} \sum_{n=0}^{\infty} [r\int_o^t \int G_s(\xi|x)dxds]^n/n!$$

$$= \exp[-r \int_o^t \int_{D_k} H_s(1-\xi|x)dxds].$$

By the monotone convergence theorem

$$G_t^{Ik}(\xi) \to G_t^I(\xi) = \exp[-r\int_o^t \int_{R^d} H_s(1-\xi|x)dxds].$$

$G_t^I(\xi)$ is shown to be a PGF by exactly the same methods as used in the proof of Theorem 5.1.1.

$$G_t^I(\xi) = \exp[\int_\eta (\exp\langle N,\log\xi\rangle-1)\Lambda_t^I(dN)]$$

where

$$\Lambda_t^I(A) = \int_o^t \int_{R^d} rP_s(A|x)dx, \quad A \in \mathcal{B}(\eta), \quad \phi \notin A \tag{5.6}$$

$$\Lambda_t^I(\{\phi\}) = 0.$$

$P_t(.|x)$ is the transition probability function for the simple branching diffusion process.

A simple computation shows that for any compact $A \subseteq R^d$,

$$\Lambda_t^I(N(A) \neq 0) \leq r|A|\left(\frac{e^{V(2\alpha-1)t}-1}{V(2\alpha-1)}\right) < \infty, \quad \alpha \neq \frac{1}{2}$$

$$r|A|t < \infty, \quad \alpha = \frac{1}{2}.$$

By Proposition 2.2, the proof is complete.

Corollary

The branching diffusion with immigration at time t is an infinitely divisible random field with KLM measure Λ_t^I.

Proposition 5.2.1

$$M^I_{(k)}(A_1\times\ldots\times A_k,t) = \sum_{j=0}^{k} \int_o^t\ldots\int_o^t \int_{(R^d)^j} r^j \sum_{\rho_k(i_1,\ldots,i_j)} \cdot$$

$$\cdot M_{(i_1)}(A_{\rho(1)}\times\ldots\times A_{\rho(i_1)},s|x_1)\ldots M_{(i_j)}(A_{\rho(k-i_j)}\times\ldots\times A_{\rho(k)},s|x_j)$$

$$\cdot dx_1\ldots dx_j ds_1\ldots ds_j \tag{5.7}$$

where " $\sum_{\rho_k(i_1,\ldots,i_j)}$ " indicates summation over all positive groupings of k distinguishable objects into j nonempty subsets.

$$M^I_{(k)}(A_1\times\ldots\times A_k,t) = \int_{A_1}\ldots\int_{A_1} \hat{P}^I_k(t,y_1,\ldots,y_k)dy_1\ldots dy_k \tag{5.8}$$

where

$$\hat{P}^I_k(t,y_1,\ldots,y_k) = \sum_{j=1}^{k} \int_o^t\ldots\int_o^t \int_{(R^d)^j} r^j \sum_{\rho_k(i_1,\ldots,i_j)}$$

$$\cdot\hat{P}_{i_1}(s,x_1,y_{\rho(1)},\ldots,y_{\rho(i_1)})\ldots\hat{P}_{i_j}(s,x_j,y_{\rho(k-i_j)},\ldots,y_{\rho(k)})$$

$$\cdot dx_1\ldots dx_j\, ds_1\ldots ds_j$$

and $\hat{P}_k(t,x,y_1,\ldots,y_k)$ is the kth order factorial moment density function of the branching diffusion with one particle initially as given in Equation (4.4).

PROOF

The proof follows by a computation similar to that of Proposition 4.1.

6. THE CENTRAL LIMIT THEOREM FOR THE BRANCHING RANDOM FIELD

By using the usual moment-cumulant formulae, factorial cumulant density functions $\hat{Q}_n$ can be obtained from the factorial moment density functions $\hat{P}_n$, when the densities exist.

Definition 6.1

(Brillinger [2]). A spatially homogeneous random field with cumulant densities $\hat{Q}_n$ is mixing if

$$\int_{(R^d)^{n-1}} \int |\hat{Q}_n(y_1,\ldots,y_{n-1},y_n)|\,dy_1\ldots dy_{n-1} < \infty$$

for $n \geq 2$.

Theorem 6.1

(Brillinger [2, Theorem 1]). Let N be a spatially homogeneous random field which is mixing. Then as $A \uparrow R^d$,

$$\frac{N(A) - E(N(A))}{\sqrt{Var(N(A))}}$$

converges in distribution to $N(0,1)$.

It is obvious that both the branching random field with uniform initial intensity and the simple branching diffusion with immigration are spatially homogeneous.

Theorem 6.2

Let $N_t(.)$ be the random field at time t associated with a branching diffusion with an initial PRF with intensity $\lambda\mu$ where μ is d-dimensional Lebesgue measure. Then as $A \uparrow R^d$

$$\frac{N_t(A) - E(N_t(A))}{\sqrt{Var(N_t(A))}} \xrightarrow{\mathcal{L}} N(0,1).$$

PROOF

By Theorem 6.1, it suffices to show that $N_t(.)$ is mixing

The cumulant densities $\hat{Q}_k$ are most easily obtained by differentiating $\log G_t(\xi)$. Let $\xi_i(x) = \chi_{A_i}(x)$, $i=1,\ldots,k$. Then if $K_{(k)}(t,A_1\times\ldots\times A_k)$ is the kth order factorial cumulant of $A_1\times\ldots\times A_k$,

$$K_{(k)}(t,A_1\times\ldots\times A_k) = \int_{A_1}\ldots\int_{A_k} \hat{Q}_k(t,y_1,\ldots,y_k)dy_1\ldots dy_k$$

or

$$K_{(k)}(t,A_1\times\ldots\times A_k) = \lim_{\eta\uparrow 1} \frac{\partial^k}{\partial\lambda_1\ldots\partial\lambda_k} \{\log G_t(\eta+\sum_1^k\lambda_i\xi_i)\}_{\lambda_1=\ldots=\lambda_k=0}.$$

From Theorem 5.1.1,

$$G_t(\xi) = \exp[\lambda\int(G_t(\xi|x)-1)dx].$$

Therefore

$$K_{(k)}(t,A_1\times\ldots\times A_k) = \lim_{\eta\uparrow 1} \frac{\partial^k}{\partial\lambda_1\ldots\partial\lambda_k}\{\lambda\int(G_t(\eta+\sum_1^k\lambda_i\xi_i|x)-1)dx\}_{\lambda_1=\ldots=\lambda_k=0}$$

$$= \lambda\int M_{(k)}(t,A_1\times\ldots\times A_k|x)dx.$$

From this it is clear that

$$\hat{Q}_k(t,y_1,\ldots,y_k) = \lambda\int\hat{P}_k(t,x;y_1,\ldots,y_k)dx \tag{6.1}$$

$$= \lambda V\alpha\int_0^t\int e^{V(2\alpha-1)u}f_k(t-u,z,y_1,\ldots,y_k)dzdu$$

from (4.4).

It must be shown that

$$\int_{(R^d)^{k-1}}\ldots\int|\hat{Q}_k(t,y_1,\ldots,y_{k-1},y_k)|dy_1\ldots dy_{k-1} \tag{6.2}$$

$$= \lambda V\alpha\int_0^t e^{V(2\alpha-1)u}\int_{R^d}\int_{(R^d)^{k-1}} f_k(t-u,z,y_1,\ldots,y_{k-1},y_k)$$

$$\cdot dy_1\ldots dy_{k-1}dz\ du$$

$$< \infty$$

But

$$\int_{(R^d)^{k-1}} f_k(t-u,z,y_1,\ldots,y_{k-1},y_k)\ dy_1\ldots dy_{k-1} \tag{6.3}$$

$$= \sum_{j=1}^{k-1}\sum_{\rho\in S_k^j}\int_{(R^d)^{k-1}}\hat{P}_j(t-u,z,y_{\rho(1)},\ldots,y_{\rho(j)})\ \cdot$$

$$\hat{P}_{k-j}(t,z,y_{\rho(j+1)},\ldots,y_{\rho(k)})\, dy_1\ldots dy_{k-1}.$$

The following integrals are evaluated by induction

$$\int_{(R^d)^j} \hat{P}_j(t,z,y_1,\ldots,y_j)dy_1\ldots dy_j \tag{6.4a}$$

$$= \begin{cases} j!e^{V(2\alpha-1)t}\left(\frac{\alpha(e^{V(2\alpha-1)t}-1}{(2\alpha-1)}\right)^{j-1}, & \alpha \neq \frac{1}{2} \\ j!\left(\frac{Vt}{2}\right)^{j-1}, & \alpha = \frac{1}{2}. \end{cases}$$

$$\int_{(R^d)^{j-1}} \hat{P}_t(t,z,y_1,\ldots,y_{j-1},y_j)dy_1\ldots dy_{j-1} \tag{6.4b}$$

$$= \begin{cases} p(t,z,y_j)j!e^{V(2\alpha-1)t}\left(\frac{\alpha(e^{V(2\alpha-1)t}-1}{(2\alpha-1)}\right)^{j-1}, & \alpha \neq \frac{1}{2}, \\ p(t,z,y_j)j!\left(\frac{Vt}{2}\right)^{j-1}, & \alpha = \frac{1}{2}. \end{cases}$$

(6.3) becomes

$$\sum_{j=1}^{k-1} \sum_{\rho\in S_k^j} p(t,z,y_j)j!(k-j)!e^{2V(2\alpha-1)(t-u)}(V\alpha m_{t-u})^{k-2}$$

$$= \quad (k-1)\binom{k}{j}\, p(t,z,y_j)j!(k-j)!\; e^{2V(2\alpha-1)(t-u)}(V\alpha m_{t-u})^{k-2}$$

where

$$m_t = \begin{cases} \frac{e^{V(2\alpha-1)t}-1}{V(2\alpha-1)}, & \alpha \neq \frac{1}{2} \\ t, & \alpha = \frac{1}{2}. \end{cases}$$

Substituting (6.3) into (6.2), and integrating,

$$\int_{(R^d)^{k-1}} |\hat{Q}_k(t,y_1,\ldots,y_{k-1},y_k)|\, dy_1\ldots dy_{k-1} = k!\lambda e^{V(2\alpha-1)t}(V\alpha m_t)^{k-1} < \infty. \tag{6.5}$$

Thus the random field is mixing and the proof is complete.

Theorem 6.3

(Ivanoff [8;Theorem 6.1]). Let $N_t^I(.)$ be the random field at time t associated with the simple branching diffusion with immigration and with no initial particles. Then as $A \uparrow R^d$,

$$\frac{N_t^I(A) - E(N_t^I(A))}{\sqrt{Var(N_t^I(A))}} \overset{\mathcal{L}}{\to} N(0,1).$$

PROOF

It suffices to show that $N_t^I(.)$ is mixing.

By the same methods as used in the proof of Theorem 6.2, it is easily seen that the kth order factorial cumulant density $\hat{Q}_k^I$ satisfies

$$\begin{aligned} \hat{Q}_k^I(t,y_1,\ldots,y_k) &= r \int_o^t \int_{R^d} \hat{P}_k(s,x,y_1,\ldots,y_k)dx\, ds \\ &= rV\alpha \int_o^t \int_o^s e^{V(2\alpha-1)u} f_k(s-u,z,y_1,\ldots,y_k)dz\, du\, ds. \end{aligned}$$

From the proof of Theorem 6.2

$$rV\alpha \int_o^s \int_{(R^d)^{k-1}} \int_{R^d} e^{V(2\alpha-1)u} f_k(s-u,z,y_1,\ldots,y_{k-1},y_k)dz\, dy_1\ldots dy_{k-1}du$$

$$= k!re^{V(2\alpha-1)s}(V\alpha m_s)^{k-1}$$

where

$$m_s = \begin{cases} \dfrac{e^{V(2\alpha-1)s}-1}{V(2\alpha-1)} & \alpha \neq \frac{1}{2} \\ s & \alpha = \frac{1}{2} \end{cases}.$$

It then follows that

$$\int_{(R^d)^{k-1}} \ldots \int |\hat{Q}_k^I(t,y_1,\ldots,y_{k-1},y_k)|dy_1\ldots dy_{k-1} = (k-1)!r(V\ )^{k-1}m_s^k < \infty. \tag{6.7}$$

Hence, the immigration process is mixing and the proof is complete.

Corollary

(Ivanoff [8, Theorem 5.4]).

(a) In the subcritical case, $\alpha < \frac{1}{2}$, the branching diffusion with immigration approaches a steady state as $t \to \infty$. The PGF of the steady state random field is

$$G_\infty^I(\xi) = \exp[r \int_{R^d} \int_0^\infty (G_u(\xi|x)-1)\, du\, dx]. \tag{6.8}$$

(b) Let $N_\infty^I(.)$ denote the steady state random field for the subcritical branching field with immigration. Then as $A \uparrow R^d$

$$\frac{N_\infty^I(A) - E(N_\infty^I(A))}{\sqrt{Var(N_\infty^I(A))}} \xrightarrow{\mathcal{L}} N(0,1).$$

PROOF

(Sketch). The proof of (a) follows by letting $t \to \infty$ in (5.5) and noting that the limiting cumulants are finite by (6.7). (b) follows in a similar way by letting $t \to \infty$ in (6.7). However we omit the details of this proof.

It should be noted here that the Central Limit Theorems 6.2 and 6.3 do not depend on the assumption that the migration process is Brownian motion. All that is needed is the Markovian property of the transition density function; i.e. that for all s,t $0 < s < t$,

$$\int_{R^d} p(t-s,a,y)p(s,y,b) = p(t,a,b).$$

7. THE CRITICAL BRANCHING RANDOM FIELD

One of the fundamental properties of the Galton-Watson branching process is that extinction occurs with probability one in the critical case $(\alpha = \frac{1}{2})$ ([1, Chapter 1]). On the other hand the effect of mixing a number of such populations is to counteract this tendency to extinction. In this section it is shown that in R^d, $d \geq 3$, that is, in the case in which the spatial Brownian motion is transient, the simple branching random field does not go to extinction but rather tends to a steady state random field.

Let $\xi(x) \equiv e^{-\varphi(x)}$ with $\varphi \in C_K(R^d)$, $\varphi \geq 0$. Let N_t denote the critical branching random field, $\alpha = 1/2$ in R^d, $d \geq 3$, with initial intensity measure $\lambda(dx) = \lambda dx$ where dx is d-dimensional Lebesgue measure. Then

$$G_t(\xi) = E(\exp(-\langle N_t, \varphi\rangle)). \tag{7.1}$$

Then by (5.1)

$$G_t(\xi) = \exp[-\lambda \int_{R^d} H_t(1-\xi|x)dx]$$

where

$$H_t(1-\xi|x) = 1-G_t(\xi|x),$$

$$0 \leq H_t(1-\xi|x) \leq 1.$$

By (4.3), $H_t(\psi|x)$ is the solution of the following Skorokhod equation

$$\begin{aligned} \frac{\partial}{\partial t}u(t,x) &= \Delta u(t,x) - \gamma u^2(t,x) \\ u(0,x) &= 1-\xi(x) \equiv \psi(x) \end{aligned} \tag{7.2}$$

where $\gamma = V/2$ and Δ is the d-dimensional Laplacian.

The proof of the existence of a steady state random field is based on an analysis of the asymptotic behavior of the solution of the S-equation (7.2) as $t \to \infty$.

Theorem 7.1

(c.f. Dawson [5, Theorem 3.1]). Let $\{N_t : t \geq 0\}$ denote the critical branching field in R^d, $d \geq 3$, with initial intensity λ. Then there is a non-degenerate limiting infinitely divisible random field which is a steady state random field for $\{N_t : t \geq 0\}$.

PROOF

The key step of the proof is to show that

$$\lim_{t\to\infty} \int H_t(\psi|x)dx = H_\infty \psi \tag{7.3}$$

exists for all $\psi \in C_K(R^d)$, $\psi \geq 0$, and that $G_\infty \xi \neq 0$ for at least one such ψ.

Then

$$G_\infty(\xi) = \exp(-\lambda H_\infty \psi), \tag{7.4}$$

$$\psi \equiv 1 - \xi$$

is the P.G.F. of a steady state random field for the simple branching random field. To verify this note that the P.G.F. of the branching random field at time $s > 0$, when the initial random field has P.G.F. G_∞, is given by

$$\begin{aligned} G_{\infty,s}(\xi) &= G_\infty(G_s(\xi|.)) \\ &= \exp(-\lambda H_\infty(H_s(1-\xi|.))) \\ &= \exp(-\lambda \lim_{t\to\infty} \int H_{t+s}(1-\xi|x)dx) \\ &= G_\infty(\xi). \end{aligned}$$

We now proceed to prove (7.3). Equation (7.2) can be rewritten as

$$H_t\psi = T_t\psi - \gamma\int_0^t T_{t-s}((H_s\psi)(H_s\psi))ds$$

where $\{T_t : t \geq 0\}$ is the semigroup defined in Section 4 with infinitesimal generator Δ.

Let $\mathfrak{T}$ denote the family of finite binary rooted trees with binary composition operation denoted by $\circ$. τ_e denotes the tree consisting of just one vertex and $\tau_1 \circ \tau_2$ denotes the rooted tree in which the root is connected to copies of τ_1, τ_2 respectively. We denote the tree $\tau_e \circ \tau_e$ by τ_s. The <u>order</u>, $|\tau|$, of $\tau \in \mathfrak{T}$ is defined as the number of boundary vertices, that is, those vertices having at most one neighbour.

The formal solution of Equation 7.5 can be written in the form

$$H_t(\psi) = T_t\psi + \sum_{\tau\in\mathcal{T}-\tau_e} (-\gamma)^{|\tau|-1} H_t^{\tau}(\psi) \tag{7.6a}$$

where $\{H_t^{\tau}(\psi):\tau\in\mathcal{T}\}$ are defined inductively as follows:

$$H_t^{\tau_e}(\psi) \equiv T_t\psi \tag{7.6b}$$

$$H_t^{\tau_1\circ\tau_2}(\psi) \equiv \int_0^t T_{t-s}(H_s^{\tau_1}(\psi)H_s^{\tau_2}(\psi))ds. \tag{7.6c}$$

It is shown below that the series (7.6a) converges absolutely thus yielding a well-defined solution provided that $\|\psi\| \equiv \sup_x|\psi(x)|$ is sufficiently small. Let

$$H_\infty^{\tau}\psi \equiv \lim_{t\to\infty}\int_{R^d} H_t^{\tau}(\psi|x)dx.$$

We now explicitly compute $H_\infty^{\tau_s}\psi$.

$$\begin{aligned} H_t^{\tau_s}(\psi|x) = \int_{R^d}\int_0^t &(2\pi(t-s))^{-d/2}\exp(-\tfrac{1}{2}|y-x|^2/(t-s)) \\ &\cdot(\int_{R^d}(2\pi s)^{-d/2}\exp(-|u-y|^2/2s)\psi(u)du) \\ &\cdot(\int_{R^d}(2\pi s)^{-d/2}\exp(-|v-y|^2/2s)\psi(v)dv)dsdy. \end{aligned} \tag{7.7}$$

Noting that

$$|u-y|^2 + |v-y|^2 = \tfrac{1}{2}|u-v|^2 + 2|y-\tfrac{1}{2}(u+v)|^2,$$

$$\begin{aligned} H_\infty^{\tau_s}\psi \equiv \lim_{t\to\infty}[\int_{R^d}\int_{R^d}\int_0^t &(2\pi(t-s))^{-\frac{d}{2}}\exp(-\tfrac{1}{2}|y-x|^2/(t-s)) \\ &\cdot(\int_{R^d}\int_{R^d}(4\pi s)^{-\frac{d}{2}}\exp(-|u-\nu|^2/4s)(\pi s)^{-\frac{d}{2}} \\ &\cdot\exp(-|y-\tfrac{1}{2}(u+\nu)|^2/s)\psi(u)\psi(\nu)dud\nu)dxdyds] \end{aligned} \tag{7.8}$$

$$= \lim_{t\to\infty} \frac{1}{2} \int_0^{2t} (2\pi s)^{-\frac{d}{2}} \left(\int_{R^d} \int_{R^d} \exp(-|u-\nu|^2/2s)\psi(u)\psi(\nu)\,du\,d\nu \right) ds$$

$$= \frac{1}{2} \int_{R^d} \int_{R^d} g(u-\nu)\psi(u)\psi(\nu)\,du\,d\nu \quad < \quad \infty.$$

The last integral is convergent since we have assumed that the Brownian motion is transient and in fact $g(.)$ is the potential kernel of the Brownian motion:

$$g(u-\nu) \quad = \quad k_d |u-\nu|^{-(d-2)} \qquad \text{where } k_d \text{ is constant.} \qquad (7.9)$$

We now obtain bounds for $H_t^{\tau}\psi$ for $\tau \in \mathcal{J}$. Since $\psi \in C_K(R^d)$, $1 \geq \psi \geq 0$, $\psi \leq \chi_K$, the indicator function of a compact set K of diameter D. Then noting that $\psi(u)\psi(u) = 0$ unless $\psi(u) > 0$ and $|w| \leq D$ where $w \equiv \nu - u$,

$$\begin{aligned} H_t^{\tau_s}(\psi|x) \quad &= \int_{R^d} \int_0^t (2\pi(t-s))^{-\frac{d}{2}} \exp(-\tfrac{1}{2}|y-x|^2/(t-s)) \\ &\cdot \left(\int_{R^d} (2\pi s)^{-\frac{d}{2}} \exp(-|u-y|^2/2s)\psi(u)du \right) \\ &\cdot \left(\int_{R^d} (2\pi s)^{-\frac{d}{2}} \exp(-|\nu-y|^2/2s)\psi(\nu)d\nu \right) ds\,dy \\ &\leq \|\psi\| \int_{R^d} \int_0^t (2\pi(t-s))^{-\frac{d}{2}} \exp(-\tfrac{1}{2}|y-x|^2/(t-s)) \\ &\cdot \left(\int_{R^d} (2\pi s)^{-\frac{d}{2}} \exp(-|u-y|^2/2s)\psi(u) \right. \\ &\cdot \left. \left(\int_{|w|\leq D} (2\pi s)^{-\frac{d}{2}} \exp(-|w+u|/2s)dw \right) du \right) ds\,dy \\ &\leq \|\psi\| T_t\psi(x) \int_0^t \int_{|w|\leq D} (4\pi s)^{-\frac{d}{2}} \exp(-|w|^2/4s)\,dw\,ds \\ &\leq \|\psi\| C(D) T_t\psi(x) \end{aligned}$$

where according to (7.9),

$$C(D) = \int_o^\infty \int_{|w|\le D} (4\pi s)^{-\frac{d}{2}} \exp(-|w|^2/4s)\,dw\,ds < \infty.$$

Using (7.6c) we obtain by induction

$$H_t^\tau(\psi|x) \le (\|\psi\|C(D))^{|\tau|-1} T_t(\psi|x)$$

and thus

$$H_\infty^\tau \psi \le (\|\psi\|C(D))^{|\tau|-2} H_\infty^{\tau_s}\psi. \qquad (7.10)$$

Hence all the limiting cumulants are finite. The tightness of the random fields $\{N_t : t \ge 0\}$ follows from the uniform bound on the second moments obtained above. It remains to show that there is a unique limiting random field and to identify its P.G.F. Note that the coefficient h_n of θ^n in the expansion of

$$H_\infty(\theta\psi) = \theta \int_{R^d} \psi(x)dx + \theta \sum_{\tau\in\mathcal{J}-\tau_e} (-\gamma\theta)^{|\tau|-1} H_\infty^\tau(\psi) \qquad (7.11)$$

satisfies $|h_n| \le C(n)k^n$, k a constant, where the power series $\Sigma_{n=1}^\infty C(n)r^n$, has a non-zero radius of convergence. In fact $C(n)$ is given by the coefficient of θ^n in the power series expansion of $f(\theta) = (1-(1-4\gamma\theta)^{1/2})/2\gamma$. Hence the limiting random field is uniquely determined by (7.11) and the proof is complete.

The key to Theorem 7.1 is the introduction of the potential kernel in the transient case. Equally it can be shown that in the recurrent case the second order moments blow up as in the Galton-Watson process, a fact first observed by S.Sawyer [16]. This suggests the possibility that extinction occurs in R^1 and R^2 for the branching random field and a more careful analysis confirms this (c.f. [5, Theorem 3.1]). In these cases J.Fleischmann [6] has obtained the following analogues of the exponential limit law (c.f. [1:Chapter 1]) for the critical Galton-Watson process.

Theorem 7.2

(Fleischmann [6:Theorems 4.1,4.2]). (a) Let N_t denote the critical branching random field in R^1. Then for A compact, $A \in \underset{\sim}{\mathcal{B}}(R^1)$

$$\lim_{t\to\infty} \mathcal{B}\sqrt{t}\; P(\frac{N_t(A)}{\alpha\sqrt{t}} > x) = F^c(x), \quad x > 0,$$

where

$$\mathcal{B} = \frac{cV}{\lambda\sqrt{2\pi}}, \qquad \alpha = \frac{V}{\sqrt{2\pi}},$$

and $F^c(.)$ is a nonincreasing function (not exponential) and c is a constant. (b) Let N_t denote the critical branching random field in R^2. Then for A compact, $A \in \underset{\sim}{\mathcal{B}}(R^2)$

$$\lim_{t\to\infty} \frac{V\log t}{\lambda 8\pi}\; P(\frac{8\pi N_t(A)}{V\mu(A)\log t} > \lambda) = e^{-x}, \qquad x > 0.$$

PROOF

(a) Let $L_t(.)$ denote the Laplace transform of the branching random field N_t. Then for $\theta > 0$, $\varphi \in C_K(R^1)$, (5.1) implies that

$$L_t(\theta\varphi) = \exp(-\lambda\int_{R^1} H_t(\psi(\theta,.)\,|x)dx)$$

where $\psi(\theta,x) \quad 1-e^{-\theta\varphi(x)}$.

Then $u(t,x) = H_t(\psi|x)$ can be written for sufficiently small θ as

$$u(t,x) = \sum_{n=1}^{\infty} \frac{(-1)^{n+1}}{n!} \int_{(R^1)^n} \hat{P}_n(t,x,y_1,\ldots,y_n)\psi(y_1)\ldots\psi(y_n)dy_1\ldots dy_n \tag{7.12}$$

where $\{\hat{P}_n\}$ are given by (4.4).

The next step is to prove by intuition that

$$\hat{P}_n(t,x,y_1,\ldots,y_n) = \hat{P}_n(1,\frac{x}{\sqrt{t}},\frac{y_1}{\sqrt{t}},\ldots,\frac{y_n}{\sqrt{t}})t^{\frac{n}{2}-1}. \tag{7.13}$$

(7.13) follows immediately for $n = 1$ since

$$\hat{P}_1(t,x,y) = (1/\sqrt{2\pi t})\exp(-(x-y)^2/2t).$$

Now assume that for $k < n$,

$$\hat{P}_k(ts,x,y_1,\ldots,y_k) = \hat{P}_k(s, \frac{x}{\sqrt{t}}, \frac{y_1}{\sqrt{t}},\ldots, \frac{y_k}{\sqrt{t}})t^{\frac{k}{2}-1}.$$

Then by (4.4)

$$\hat{P}_n(t,x,y_1,\ldots,y_n)$$

$$= \gamma \sum_{\rho\in S_n^k} \sum_{k=1}^{n-1} \frac{1}{k!}\frac{1}{(n-k)!} \int_0^t \int p(t-s,x,y)\hat{P}_k(s,y,y_{\rho(1)},\ldots,y_{\rho(k)})$$

$$\cdot \hat{P}_{n-k}(x,y,y_{\rho(k+1)},\ldots,y_{\rho(n)})dsdy$$

$$= \gamma \sum_{\rho\in S_n^k} \sum_{k=1}^{n-1} \frac{1}{k!}\frac{1}{(n-k)!} t^{\frac{3}{2}} \int_0^1 \int p(t-st,x,y\sqrt{t})$$

$$\cdot \hat{P}_k(st,y\sqrt{t},y_{\rho(1)},\ldots,y_{\rho(k)})P_{n-k}(st,y\sqrt{t},y_{\rho(k+1)},\ldots,y_{\rho(n)})dsdy$$

$$= \gamma \sum_{\rho\in S_n^k} \sum_{k=1}^{n-1} \frac{1}{k!}\frac{1}{(n-k)!} t^{\frac{n}{2}-1} \int_0^1\int p(1-s,\frac{x}{\sqrt{t}},y)$$

$$\cdot\ \hat{P}_k(s,y,\frac{y_{\rho(1)}}{\sqrt{t}},\ldots,\frac{y_{\rho(k)}}{\sqrt{t}})\ \hat{P}_{n-k}(s,y,\frac{y_{\rho(k+1)}}{\sqrt{t}},\ldots,\frac{y_{\rho(n)}}{\sqrt{t}})dsdy$$

$$= \hat{P}_n(1,\frac{x}{\sqrt{t}}, \frac{y_1}{\sqrt{t}},\ldots, \frac{y_n}{\sqrt{t}})t^{\frac{n}{2}-1}.$$

Hence

$$u(t,x) = \sum_{n=1}^{\infty} (-1)^{n+1} \frac{t^{\frac{n}{2}-1}}{n!} \int P_n(1,\frac{x}{\sqrt{t}}, \frac{y_1}{\sqrt{t}},\ldots, \frac{y_n}{\sqrt{t}}) \tag{7.14}$$

$$\cdot\psi(y_1)\ldots\psi(y_n)\ dy_1\ldots dy_n.$$

Therefore

$$L_t(\frac{\theta\varphi}{\alpha\sqrt{t}}) = \exp(\lambda \sum_{n=1}^{\infty} (-1)^n \frac{t^{\frac{n}{2}-1}}{n!} \int_{R^1} \int_{(R^1)^n} \hat{P}_n(1, \frac{x}{\sqrt{t}}, \frac{y_1}{\sqrt{t}},\ldots, \frac{y_n}{\sqrt{t}})$$

$$\cdot \prod_{i=1}^{n} (1-e^{\frac{-\theta\varphi(y_i)}{\alpha\sqrt{t}}})dy_1\ldots dy_n\ dx).$$

Hence using Taylor's remainder theorem and the continuity of the functions $\hat{P}_n(1,x,.,.,.,.,.)$ for the branching diffusion in R^1,

$$L_t\left(\frac{\theta\varphi}{\alpha\sqrt{t}}\right) = \exp\left[\lambda \sum_{n=1}^{\infty} (-1)^n t^{-\frac{1}{2}n}\theta^n \alpha^{-n}\ell(n) + t^{-1}O\left(\sum_{n=1}^{\infty} (-\theta^2)^n k^n r_n(\theta)\ell(n)\right)\right] \tag{7.15}$$

where $k \ge 0$, $0 \le r_n(\theta) \le 1$, and

$$\ell(n) = \left(\int\varphi(y)dy\right)^n \left(\int_{R^1} P_n(1,x,0,\ldots,0)dx\right)(n!)^{-1}.$$

Using an argument similar to that of the proof of (7.10) it can be shown that

$$\int_{R^1}\int_{(R^1)^n} \hat{P}_n(t,x,y_1,\ldots,y_n)\psi(y_1)\ldots\psi(y_n)dy_1\ldots dy_n dx \le n!C(n)t^{\frac{n-1}{2}} k^n$$

for a positive constant k. But in the proof of Theorem 7.1 it was shown that the power series $\Sigma_{n=1}^{\infty} C(n)r^n$ has a non-zero radius of convergence . Let $F_t(.)$ denote the cumulative distribution function of $\langle N_t,\varphi\rangle/\alpha\sqrt{t}$. Then

$$\int_0^{\infty} e^{-\theta x}dF_t(x) = \exp(G(\theta)/\sqrt{t} + t^{-1}g(\theta))$$

where

$$G(\theta) = \lambda \sum_{n=1}^{\infty} (-1)^n\theta^n\ell(n)$$

and $g(\theta)$ is bounded for bounded sets of θ. Letting $F_t^c(x) \equiv 1-F_t(x)$ and using integration by parts,

$$\sqrt{t}\int_0^{\infty} e^{-\theta x}F_t^c(x)dx = -\frac{\sqrt{t}}{\theta}\left[\exp\left(\frac{G(\theta)}{\theta} + t^{-1}g(\theta)\right)-1\right].$$

Hence

$$\sqrt{t}\int_0^{\infty} F_t^c(x)dx \le \lim_{\theta\downarrow 0} \sqrt{t}\int_0^{\infty} e^{-\theta x}F_t^c(x)dx$$

$$= \frac{\lambda\ell(1)}{\alpha} < \infty .$$

But (7.15) also yields a uniform bound on the second moments thus guaranteeing tightness of the measures $\mathcal{B}t^{1/2}F_t^c(x)dx$ on $[0,\infty)$.

It remains to show that the limit is unique, but this follows since for sufficiently small $\theta > 0$,

$$\mathcal{B}L(\theta) \equiv \lim_{t\to\infty} \mathcal{B}\sqrt{t} \int_0^\infty e^{-\theta x} F_t^c(x)dx = \mathcal{B}\lambda \sum_{n=1}^\infty (-1)^{n+1}\alpha^{-n}\theta^{n-1}(\int\varphi(y)dy)^n \ell^1(n)$$

where

$$\ell^1(n) = (\int_{R^1} \hat{P}_n(1,x,0,\dots,0)dx)(n!)^{-1}.$$

Hence $\mathcal{B}t^{1/2}F_t^c(.) \to F^c(x)$ where

$$-\int_0^\infty x^n dF^c(x) = \mathcal{B}\lambda\alpha^{-n}\ell(n)n!, \qquad n \geq 1,$$

and the proof is complete except for the routine calculation of α and $\mathcal{B}$.

By computing the first few moments, Fleischmann [6] showed that this "distribution" $F^c(.)$, is not gamma, Mittag-Lefler or exponential. Note moreover that the above proof does not imply that $\lim_{x\downarrow 0} F^c(x) < \infty$!

The proof of (b) is based on the estimate

$$E_x((N_t(A))^n) = \frac{n!(\log t)^{n-1}(\lambda\mu(a))^n V^{n-1}}{(2\pi)^n 4^{n-1} t} \exp(\frac{-(x-c_x)^2}{2t})$$
$$+ O(\frac{c^n n!}{t}(\log t)^{n-2}) \exp(\frac{-(x-c_x^1)^2}{2t}),$$

$t \geq e$, c_x is defined so that $(x-c_x)^2 = \min_{c\in A}(x-c)^2$, c_x^1 is defined so that $(x-c_x^1)^2 = \min_{c\in A+r(A)}(x-c)^2$, and $r(A)$ is 1/2 the length of the diameter of A. However we omit the proof and refer the reader to [6] for details.

8. THE FUNCTIONAL CENTRAL LIMIT THEOREM FOR THE CRITICAL STEADY STATE

In the previous section it was shown that a critical branching random field in R^d, $d \geq 3$, has a steady state random field which we denote by N_∞. In this section we investigate this steady state random field and derive a functional central limit theorem for it. Note that in this case Brillinger's theorem does not apply since the random field is not mixing.

In order to formulate the functional central limit theorem it is necessary to introduce the renormalization transformation. Given $K > 0$ and $\psi \in C_K(R^d)$, let

$$\psi_K(x) \equiv \psi(x/K). \tag{8.1}$$

Consider the transformed random measure N_∞^K defined by

$$\langle N_\infty^K,\psi\rangle \equiv \langle N_\infty,\psi_K\rangle = \int\psi(x/K)N_\infty(dx). \tag{8.2}$$

The effect of this transformation is to reduce the spatial dimensions by a factor of K, that is, the random measure assigned to the unit cube by N_∞^K equals the random measure assigned to the cube of length K by the N_∞ random field.

The functional central limit theorem consists of finding constants b_K such that $(N_\infty^K-E(N_\infty^K))/b_K$ converges in distribution to a limiting random field as $K \to \infty$. In the case under consideration the limiting random field is a Gaussian generalized random field with mean zero and is therefore characterized by its covariance kernel.

Theorem 8.1

(c.f. Dawson [5;Theorem 5.1]). Let N_∞ denote the steady state random field for the critical branching random field in R^d, $d \geq 3$. Then

$$(N_\infty^K-E(N_\infty^K))/K^{(d+2)/2}$$

converges in distribution to a Gaussian generalized random field with mean zero and covariance kernel $g(.)$, the potential kernel of the d-dimensional Brownian motion.

PROOF

Let $L_\infty^K(.)$ denote the Laplace transform of the random field

N_∞^K. Then for $\theta > 0$, $\varphi \in C_K(R^d)$, $\varphi \geq 0$,

$$L_\infty^K(\theta\varphi_K) = \exp(-\lambda H_\infty \psi_K(\theta,.)) \tag{8.3}$$

where

$$\psi(\theta,x) \equiv 1 - e^{-\theta\varphi(x)}$$

and

$$H_\infty\psi(\theta,.) = \lim_{t\to\infty} \int_{R^d} H_t(\psi|x)dx.$$

If $u^{(K)}(t,x) \equiv H_t(\beta\psi_K|x), \beta > 0$, then $u^{(K)}(t,x)$ is the solution of the S-equation

$$u^{(K)}(t,x) = \beta T_t\psi - \gamma\int_0^t T_{t-s}(u^{(K)}(s,.))^2 ds. \tag{8.4}$$

In order to study the limit we must investigate the dependance of the solution of this integral equation on K. We do this by writing the solution as a power series in θ and then determining the dependance of the coefficients on K. The result is described in the following lemma.

Lemma 8.1

For sufficiently small β,

(a) $$u^{(K)}(t,x) = \sum_{k=1}^{\infty} \beta^k u_{\infty,k}(t/K^2, x/K)K^{2(k-1)}$$

($u_{\infty,k}(.,.)$ is a function which will be defined in the proof).

(b) In terms of the factorial moment density functions, this is equivalent to

$$\hat{P}_k(t,x;y_1,\ldots,y_k) = K^{k(2-d)-2}\hat{P}_k\left(\frac{t}{K^2},\frac{x}{K};\frac{y_1}{K},\ldots,\frac{y_k}{K}\right).$$

PROOF

Let

$$u_1^{(K)}(t,x) \equiv (2\pi t)^{-d/2}\beta\int_{R^d} \exp(-\tfrac{1}{2}|y-x|^2/t)\psi(\theta,\tfrac{y}{K})dy \tag{8.5}$$

$$= \theta(2\pi t/K^2)^{-\frac{d}{2}} \int_{R^d} \exp(-\frac{1}{2}|\frac{y}{K}-\frac{x}{K}|^2/tK^2)\psi(\theta,\frac{y}{K})K^{-d}dy.$$

Hence

$$u_1^{(K)}(t,x) = u_1(t/K^2,x/K). \tag{8.6}$$

Let $u_n^{(K)}(.,.)$ be defined recursively by

$$u_{n+1}^{(K)}(t,x) \equiv u_1^{(K)}(t,x) - \gamma\int_0^t \int_{R^d} (2\pi(t-s))^{-\frac{d}{2}}\exp(-\frac{1}{2}|y-x|^2/(t-s))$$

$$\cdot [u_n^{(K)}(s,y)]^2 dyds \tag{8.7}$$

and $u_{n+1}(t,x) \equiv u_{n+1}^{(1)}(t,x)$.

As in the proof of Theorem 7.1 we note that for sufficiently small θ, $\{u_n^{(K)}\}$ converges to the unique solution of (8.4). Note that

$$u_n(s,y) = \sum_k \theta^k u_{n,k}(s,y) \tag{8.8}$$

where only a finite number of terms are non-zero. The next step is to prove by mathematical induction that

$$u_n^{(K)}(s,y) = \sum_k \theta^k u_{n,k}(s/K^2,y/K)K^{2k-2}. \tag{8.9}$$

(8.9) has been proved in the case $n=1$ in (8.6). To carry out the induction step $n \to n+1$, recall that by definition

$$u_{n+1}^{(K)}(t,x) = u_1(t/K^2,x/K) - \gamma K^2\int_0^t(\int_{R^d} (2\pi(t-s)/K^2)^{-\frac{d}{2}}$$

$$\cdot \exp(-\frac{1}{2}|\frac{y}{K} - \frac{x}{K}|^2/((t-s)/K^2))[u_n^{(K)}(s,y)]^2K^{-d}dy)K^{-2}ds$$

$$= u_1(t/K^2,x/K) - \gamma K^2 \int_0^{t/K^2} \int_{R^d} (2\pi(\frac{t}{K^2}-r))^{-\frac{d}{2}}$$

$$\cdot\exp(-\frac{1}{2}|u- \frac{x}{K}|^2/(t/K^2-r))[\sum_k K^{2k-2}\theta^k u_{n,k}(r,u)]^2 dudr$$

$$= \sum_k \theta^k u_{n+1,k}(t/K^2,x/K)K^{2k-2}.$$

Hence the proof of (8.9) by induction is complete since for sufficiently small β,

$$u^{(K)}(t,x) \equiv \lim_{n\to\infty} u_n^{(K)}(t,x),$$

and if $\lim_{n\to\infty} u_{n,k} \equiv u_{\infty,k}$ then

$$u^{(K)}(t,x) = \sum_{k=1}^{\infty} \beta^k u_{\infty,k}(t/K^2, x/K) K^{2(k-1)}$$

and the proof of (a) is complete.

To prove (b), notice that

$$\begin{aligned} u^{(K)}(t,x) &= H_t(\beta\psi_K | x) \\ &= \sum_{k=1}^{\infty} \beta^k \frac{(-1)^{k+1}}{k!} \int_{(R^d)^k} \hat{P}_k(t,x,y_1,\ldots,y_k) \\ &\quad \cdot \psi(\tfrac{y_1}{K}) \ldots \psi(\tfrac{y_k}{K})\, dy_1 \ldots dy_k \\ &= \sum_{k=1}^{\infty} \beta^k u_k(\tfrac{t}{K2}, \tfrac{x}{K}) K^{2(k-1)} \qquad \text{by part (a).} \end{aligned}$$

Therefore,

$$u_k(\tfrac{t}{K^2}, \tfrac{x}{K}) K^{2(k-1)} = \frac{(-1)^{k+1}}{k!} K^{kd} \int_{(R^d)^k} \hat{P}_k(t,x,Kz_1,\ldots,Kz_k) \cdot \psi(z_1)\ldots\psi(z_k)\, dz_1\ldots dz_k.$$

But for $K = 1$,

$$u_k(t,x) = \frac{(-1)^{k+1}}{k!} \int_{(R^d)^k} \hat{P}_k(t,x,y_1,\ldots,y_k)\psi(y_1)\ldots\psi(y_k)dy_1\ldots dy_k,$$

and so

$$u_k(\tfrac{t}{K^2}, \tfrac{x}{K}) K^{2(k-1)} = \frac{(-1)^{k+1}}{k!} K^{2(k-1)} \int_{(R^d)^k} \hat{P}_k(\tfrac{t}{K^2}, \tfrac{x}{K}, y_1,\ldots,y_k) \cdot \psi(y_1)\ldots\psi(y_k)\, dy_1\ldots dy_k.$$

The result follows immediately.

We now return to the proof of Theorem 8.1. Using part (a) of

Lemma 8.1,

$$\int_{R^d} u^{(K)}(t,x)dx = \int_{R^d} \Big(\sum_{k=1}^{\infty} u_{\infty,k}(t/K^2, x/K) K^{2(k-1)} \beta^k \Big) dx$$

$$= \int_{R^d} \sum_{k=1}^{\infty} u_{\infty,k}(t/K^2, u) K^{2(k-1)+d} \beta^k) du$$

and hence

$$\lim_{t\to\infty} \int_{R^d} u^{(K)}(t,x)dx = \sum_{k=1}^{\infty} [\lim_{t\to\infty} \int_{R^d} u_{\infty,k}(t,u)du] K^{2(k-1)+d} \beta^k.$$

The fact that this limit is finite follows from (7.10). Thus we have obtained the cumulant generating functional

$$\Phi_K(\theta\varphi) \equiv \log E(\exp(-\langle N_\infty - E(N_\infty), \theta\varphi_K\rangle))$$

$$= \sum_{k=2}^{\infty} K^{2(k-1)+d} \beta^k H_{\infty,k} \psi(\theta,.)\Big|_{\beta=1}$$

where

$$H_{\infty,k}\psi(\theta,.) = \lim_{t\to\infty} \int_{R^d} u_{\infty,k}(t,u)\, du.$$

In order to compute $\Phi_K(\theta\varphi_K/K^{(d+2)/2})$, note that for large K,

$$\psi\Big(\frac{\theta}{K^{(d+2)/2}}, x\Big) = \theta K^{-(d+2)/2}\varphi(x) + \varepsilon(x,K)\varphi(x)$$

and

$$\sup_x |\varepsilon(x,K)| K^{(d+2)/2} \to 0 \quad \text{as} \quad K \to \infty.$$

Thus the cumulant generating functional has the following form, where the a_k's are constants not depending on K, θ, or β:

$$\Phi_K(\theta\varphi/K^{(d+2)/2}) = \sum_{k=2}^{\infty} a_k K^{(d-2)(1-\frac{1}{2}k)} \theta^k + o(\theta,K)$$

where $o(\theta,K) \to 0$ as $K \to \infty$ uniformly for bounded sets of θ. Hence if $d > 2$,

$$\lim_{K\to\infty} \Phi_K(\theta\varphi/K^{(d+2)/2}) = a_2\theta^2,$$

the cumulant generating functional of a Gaussian generalized random field. The covariance kernel is given by

$$H_{\infty,2}\varphi = \frac{1}{2}\gamma\lambda \int_{R^d}\int_{R^d} g(u-v)\varphi(u)\varphi(v)\,du\,dv$$

by (7.8) and the proof is complete.

Remark 8.1

In Section 6 central limit theorems were obtained for the branching random field at a finite time and for the branching random field with immigration. The results of that section could be extended to the functional form but would have the Dirac delta function as covariance kernel, that is, unlike the case treated in this section the limiting random field would give independent values to disjoint sets.

9. THE DIFFUSION APPROXIMATION

W. Feller first introduced the idea of using the diffusion approximation to study stochastic population models. Recently the diffusion approximation has been extended to spatially distributed stochastic population models ([4],[5]). The diffusion in this case is a Markov measure-valued process with state space $(\mathcal{M}, \underset{\sim}{B}(\mathcal{M}))$. In this section we give an elementary derivation of the small particle limit or diffusion approximation to the branching random field.

The basic idea is to consider each particle to have a mass and to consider the situation in which there are a large number of small particles.

For each $K > 0$, let:

mass of each particle $= m_K = \frac{1}{K}$

initial intensity $= \lambda_K = K\lambda$

branching rate $= V_K = KV$

Define $N_T^K \equiv \frac{1}{K}N_t$.

Then if $\xi(x) = e^{-\varphi(x)}$

$$KH_0^K(1-\xi|x) = K(1-e^{-\varphi(x)/K}),$$

$$H_t^K(1-\xi|x) = 1-E(\exp(-\langle N_t^K,\varphi\rangle)|\{x\}).$$

It is easy to verify that $KH_t^K(1-\xi|x)$ also satisfies Equation (7.2). Thus

$$KH_t^K(1-\xi|x) = H_t(K(1-e^{-\frac{\varphi}{K}})|x).$$

Thus

$$L_{N_t^K}(\varphi) = \exp(-\lambda\int_{R^d} H_t(K(1-e^{-\frac{\varphi}{K}})|x)dx).$$

Then

$$L_t(\varphi) = \lim_{K\to\infty} L_{N_t^K}(\varphi)$$

$$= \lim_{K\to\infty} \exp(-\lambda \int_{R^d} H_t(K(1-e^{-\frac{\varphi(x)}{K}})|x)dx).$$

Hence

$$L_t(\varphi) = \exp(-\lambda \int_{R^d} H_t(\varphi|x)dx).$$

$L_t(\varphi)$ is the Laplace transform of a multiplicative measure-valued Markov process which is called a measure diffusion process [4]. The multiplicative measure diffusion process is an example of a branching process with measure-valued states which were introduced by Jirina [10]. Measure diffusions have recently been studied by Dawson [4],[5]. Hence we have

Theorem 9.1

The limit of the measure-valued processes N_t^K is a measure valued Markov process $\{X_t:t \geq 0\}$ with Laplace transform

$$L_{X_t}(\varphi) = \exp(-\lambda \int_{R^d} H_t(\varphi|x)dx). \quad (9.1)$$

The following theorem shows that there is an intimate connection between the simple branching random field and the multiplicative measure diffusion process.

Theorem 9.2

Let $\{X_t : t \geq 0\}$ denote the multiplicative measure diffusion process with $X_o = \lambda\mu$ where μ is Lebesgue measure and $\lambda > 0$. Let $\{N_t : t \geq 0\}$ denote the simple branching random field with initial intensity measure $\lambda\mu$. Then for each $t > 0$, N_t has the same distribution as the doubly stochastic Poisson random field in which the random intensity is X_t.

PROOF

Let Y_t denote the doubly stochastic Poisson random field with random intensity X_t. Then

$$
\begin{aligned}
L_{Y_t}(\varphi) &= L_{X_t}(1-e^{-\varphi}) && \text{(by 2.7)} \\
&= \exp(-\lambda \int_{R^d} H_t(1-e^{-\varphi}|x)dx) && \text{(by 9.1)} \\
&= L_{N_t}(\varphi) && \text{(by 5.1)}
\end{aligned}
$$

and the proof is complete.

Remark 9.1

Although the S-semigroups become more complex for more general branching mechanisms, the above diffusion approximation is universal for the family of branching mechanisms in which the offspring distribution has finite second moment and the offspring are located at the location of the parent. It is because of this analytical simplicity that the diffusion approximation is such a useful tool.

10. RELATED DEVELOPMENTS

A <u>critical cluster process</u> is a discrete time multiplicative Markov process whose value at each time is a random field. The evolution takes place by replacing each particle at one epoch by a random cluster of particles distributed in space at the next epoch.

It is assumed that the expected number of offspring is one. The process is said to be stable if starting with a Poisson random field this process converges in distribution to a nontrivial random field which is invariant under the clustering operation. The stability of cluster processes has been studied in detail by J. Kerstan, K. Matthes and J.Mecke [11]. The branching random field can be viewed as a continuous time version of a cluster process and the multiplicative measure diffusion is the analogue when random fields are replaced by arbitrary random measures.

The results described in this paper for the simple branching random field can be generalized to more general branching mechanism and motion processes. For example in Dawson [5] the results are extended to the case in which the motion process is a symmetric stable process. Once again in this case the existence of a steady state coincides with the class of transient symmetric stable processes and the functional central limit theorem leads to a Gaussian generalized random field whose covariance kernel is the corresponding Riesz potential kernel.

Finally, it would clearly be desirable to study such systems in which there are interactions between the particles. However completely new methods are required and results have only been obtained in a few special cases. T.M.Liggett [12] provides an excellent introduction to the study of interacting particle systems when the particles are constrained to live on the lattice.

REFERENCES

1. K.Athreya and P.Ney, Branching Processes, Springer Verlag, New York (1972).
2. D.R.Brillinger, In Stochastic Processes and Related Topics (M.L.Puri, ed.) (1974).
3. D.J.Daley and D.Verre-Jones, In Stochastic Point Processes: Statistical Analysis, Theory and Applications (P.A.W.Lewis,ed.) Wiley, New York (1972).

4. D.A.Dawson, J. Mult.Anal. 5:1 (1975).
5. D.A.Dawson, Z.Wahrscheinlichkeitsth. (to appear).
6. J.Fleischmann, Limiting Distributions for Critical Branching Brownian Random Fields, Yeshiva University (1976).
7. N.Ikeda, M.Nagasawa and S.Watanabe, J.Math.Kyoto Univ. 8:233 and 365, 9:95 (1968-69).
8. G. Ivanoff, Branching Diffusion Processes, PH.D. Thesis, Carleton University (1976).
9. P.Jagers, Advances in Probability 3:179 (1974).
10. M. Jirina, Trans.Third Prague Conf.Inform.Theory, Statist. Decision Functions, Random Processes, Prague (1964).
11. J.Kerstan, K.Matthes and J.Mecke, Unbregrenzt teilbare Punktprozess. Akademie-Verlag, Berlin (1974).
12. T.M.Liggett, Lecture Notes in Mathematics, Springer Verlag (to appear).
13. J.E.Moyal, Acta Math.98:221 (1957),
14. J.E.Moyal, Acta Math. 108:1 (1962).
15. J.E.Moyal, J.Applied Prob.1:267 (1964).
16. S.Sawyer, Adv.Appl.Prob.8:650 (1976).
17. A.Skorokhod, Theor.Prob.Appl.9:492 (1964).
18. M.Westcott, J.Austral.Math.Soc. 14:448 (1972).

STATISTICAL INFERENCE IN BRANCHING PROCESSES

Jean-Pierre Dion

Département de Mathématiques
Université du Québec à Montréal
Montreal, P.Q., Canada

and

Niels Keiding

Institute of Mathematical Statistics
University of Copenhagen
Copenhagen, Denmark

INTRODUCTION

This paper surveys the theoretical literature on statistical inference for branching processes, almost exclusively from the last five years. The emphasis is on estimation of the offspring mean (time-discrete case) and the Malthusian growth parameter (time-continuous case) based on a complete record of the process in a time period.

Essentially two different repetitive structures are important: increasing number of ancestors, leading to standard iid theory, and increasing number of generations with which we shall mostly be concerned. Very little thought has been given to small-sample theory, including suitable conditionality arguments for defining reasonable reference sampling distributions.

For the large sample theory for supercritical branching processes in discrete time it seems useful to view a series of generations as successive sections (of random length) of iid replications of a random variable with the offspring distribution. The continuous-time analogue is the random time transformation to a Poisson process (with has iid increments) via the inverse of the natural increasing process corresponding to the split time process, this being a trivial submartingale. The latter approach has been used by Aalen for statistical analysis of more general point processes.

The use of the natural increasing process as a measure of efficiency has been studied for both discrete and continuous time by a series of Australian authors, who have also provided suggestions as to definition of efficiency concerning tests of hypotheses on the offspring mean. Essentially, these concepts reduce consideration to the underlying structure of iid birth events, but the general formulation suggests wider applicability within the theory of statistical inference for stochastic processes.

The large sample results for the estimators state strong consistency on the set of nonextinction and asymptotic normality, using the above mentioned natural increasing process (total exposure time) as random normalising factor. A unifying tool is martingale central limit theory. Other areas are the subcritical process with immigration, for which some time-series analogue estimators have been investigated by martingale methods. Very recently, also multitype processes have been studied. For this case as well as for Bellman - Harris processes in continuous time, the rate of convergence of the estimator towards the parameter turns out to be critically dependent on the rate of convergence to the stable type (or age) distribution.

The paper is divided into four chapters. Estimation theory is discussed in Chapters 1 (discrete time) and 2 (continuous time). A separate exposition of the role of martingale theory appears in Chapter 3 and Chapter 4 provides a short general discussion of the possible applicability of the theory as well as two examples of applications.

1. ESTIMATION THEORY FOR DISCRETE-TIME BRANCHING PROCESSES

1.1 Maximum likelihood estimation of the parameters of the offspring distribution.

It is basic for the statistical theory for branching processes to exploit the underlying structure of independent identically distributed offspring sizes. We shall therefore define the Galton-Watson process by first considering a sequence $Y_1, Y_2, \ldots$ of iid random variables on $\mathbb{N}=\{0,1,2,\ldots\}$, all distributed according to the <u>offspring distribution</u> (p_k). Let z_0 be some fixed number and define the stochastic process $Z_0, Z_1, \ldots$ by

$$Z_0 = z_0 ,$$

$$Z_1 = Y_1 + \ldots + Y_{Z_0} ,$$

$$Z_2 = Y_{Z_0+1} + \ldots + Y_{Z_0+Z_1} ,$$

$$\ldots$$

$$Z_n = Y_{Z_0+\ldots+Z_{n-2}+1} + \ldots + Y_{Z_0+\ldots+Z_{n-1}} ,$$

$$\ldots$$

Then (Z_n) is a Galton-Watson process with offspring distribution (p_k) and $Z_0 = z_0$.

<u>Proposition 1.1</u> Consider the completely general statistical model specified by all offspring distributions $\{(p_k):\Sigma p_k = 1\}$ and assume that all individual offspring sizes of individuals in the n first generations have been observed. Then the maximum likelihood estimator of the (p_k) is the obvious set of relative frequencies

$$\hat{p}_k = N_k/(Z_0+\ldots+Z_{n-1})$$

where N_k is the number of times k offspring are produced, that is

$$N_k = \#\{i \in \{1,\ldots,Z_0+\ldots+Z_{n-1}\} \mid Y_i = k\}.$$

<u>Proof</u>. If $Y_1,\ldots,Y_p$ had been observed, obviously

$$\hat{p}_k = \# \{i \in \{1,\ldots,p\} \mid Y_i = k\}/p.$$

In the present context, p has been replaced by the <u>stopping time</u> $Z_0+\ldots+Z_{n-1}$ which does not depend on the parameters. At the point $(y_1,\ldots,y_{z_0+\ldots+z_{n-1}})$ the likelihood of $(Y_1,\ldots,Y_{Z_0+\ldots+Z_{n-1}})$ is proportional to that of $(Y_1,\ldots,Y_p)$ with $p = z_0+\ldots+z_{n-1}$ (this property is conventionally termed "the likelihood is independent of the stopping rule"). This concludes the proof. The result was first pointed out by Harris [31].

Corollary. The maximum likelihood estimator of the offspring mean $m = \Sigma k p_k$ in the general model based on the observation of all offspring sizes in the n first generations is given by

$$\hat{m} = (Z_1+\ldots+Z_n)/(Z_0+\ldots+Z_{n-1}),$$

that is, the total number of children divided by the total number of parents.

Proof. $\hat{m} = \Sigma k\hat{p}_k$.

Based on the result of the Corollary, it is an obvious conjecture that $\hat{m}$ is also the maximum likelihood estimator in the general model if only the generation sizes $Z_0,\ldots,Z_n$ are observed.

Theorem 1.2 The maximum likelihood estimator of the offspring mean m in the general model based on the observation of the generation sizes $Z_0,\ldots,Z_n$ only is given by

$$\hat{m} = (Z_1+\ldots+Z_n)/(Z_0+\ldots+Z_{n-1}).$$

Proof. Dion [21, Théorème 1.2] and Feigin [28] gave Lagrange multiplier arguments. Keiding and Lauritzen [54] used an exponential family approach which also delineates a class of statistical models in which $\hat{m}$ is the maximum likelihood estimator of the offspring mean, as shown below.

Remark. Notice that $\hat{m}$ is always finite even though the model allows for $m = \infty$.

Theorem 1.3 (Keiding and Lauritzen [54]). Let the parameter set $\Theta \subseteq R\times\Omega$, $\theta = (\xi,\omega) \in \Theta$ and the offspring probabilities be of the form

$$p_x(\theta) = a(\theta)g(x,\omega)e^{\xi x}$$

where the support $S = \{x \mid g(x,\omega) > 0\}$ is independent of θ and for each fixed ω the sections $\Xi_\omega = \{\xi \in R^k \mid (\xi,\omega) \in \Theta\}$ are open and given as

$$\Xi_\omega = \{\xi \in R^k \mid \sum_{x\in S} g(x,\omega)e^{\xi x} < \infty\}.$$

Then if the observed $\hat{m}$ belongs to the interior of the closed convex hull of S, it is the maximum likelihood estimate of m.

Corollary. If N is a stopping time not depending on the parameters, and if $Z_0,\ldots,Z_N$ is observed, then the maximum likelihood estimator of m in the models described in the previous theorems is given by

$$(Z_1+\ldots+Z_N)/(Z_0+\ldots+Z_{N-1}).$$

Examples of offspring distributions fulfilling the conditions of Theorem 1.3 are the power series distributions (Dion [21], Becker [11], Eschenbach and Winkler [26], Heyde [35], Heyde and Feigin [36], binary splitting (Jagers [46], the modified geometric distribution (Keiding [50]) and the negative binomial distribution. A nontrivial counterexample is the zeta-distributions with probabilities $p_x = (x + 1)^{-\theta}/\zeta(\theta)$, $x = 0,1,2,\ldots$, $1 < \theta < \infty$, with ζ the Riemann zeta function, for which $\hat{m}$ is not the maximum likelihood estimator (Keiding [50]).

Observation of a random number N of generation sizes has in particular been studied by Becker [11].

We conclude this section by noticing that there exist practically no small-sample distribution theory for $\hat{m}$, and the interesting question of possible conditional inference has hardly been touched upon.

1.2 Asymptotic theory for $\hat{m}$.

Essentially two different repetitive structures are important: increasing number of ancestors and increasing number of generations.

By the branching property, the distribution of $(Z_0,\ldots,Z_N)$ given $Z_0 = z$ is the same as that of a sum of z independent identically distributed replications of $(Z_0,\ldots,Z_n)$ given $Z_0 = 1$. It follows that standard large sample theory applies for large z, the most complete and careful treatment being that of Yanev [72].

Theorem 2.1 For fixed n and $z \to \infty$,

(a) if $m \to \infty$, then $\hat{m} \to m$ a.s. and $E(\hat{m}) \to m$,

(b) if the offspring variance $\sigma^2 < \infty$, then the asymptotic distribution of

$$\{\frac{z}{\sigma^2}(1+m+\ldots+m^{n-1})\}^{\frac{1}{2}}(\hat{m}-m)$$

is standard normal.

Remark. An iterated logarithm result is routine.

We next consider the second asymptotic structure, that of increasing number of generations, first with fixed number z of ancestors, conveniently assumed equal to one. Since a branching process has all states $j > 0$ transient, there is no way of appealing to the standard theory of statistical inference for Markov chains such as given by Billingsley [15]. Obviously, nontrivial results exist only for the supercritical case $m > 1$; we let A denote the set $\{Z_n \to \infty\}$ of nonextinction.

Theorem 2.2 If $1 < m < \infty$, then for fixed z and $n \to \infty$

$\hat{m} \to m$ a.s. on A.

Proof. This result was noted by Heyde [33] as a corollary to his definitive theorem on normalizing constants for the asymptotic growth of supercritical branching processes. We take here the opportunity to point out that the result is in fact an easy consequence of the strong law of large numbers in its simplest form.

Let $Y_1, Y_2, \ldots$ be the iid offspring sizes defined in Section 1. Then as $p \to \infty$, $(Y_1+\ldots+Y_p)/p \to m$ a.s. On the set $\{Z_0+\ldots+Z_{n-1} \to \infty\}$, which is a.s. the same as A, it follows that

$$\hat{m} = \frac{Y_1+\ldots+Y_{Z_0+\ldots+Z_{n-1}}}{Z_0+\ldots+Z_{n-1}} \to m$$

a.s.

Asymptotic normality is obtained in a similar manner.

Theorem 2.3 (Dion [21,22], Jagers [44]). Assume $m > 1$, $0 < \sigma^2 < \infty$ and let $P_A\{\cdot\} = P\{\cdot \mid A\}$ and $S(w) = P_A\{W \leq w\}$ where as usual $W = \lim_{n\to\infty}$ a.s. $Z_n m^{-n}$. Then for fixed z and $n \to \infty$

(a) $P_A\{\sigma^{-1}(Z_0+\ldots+Z_{n-1})^{\frac{1}{2}}(\hat{m}-m) \leq x\} \to \Phi(x) = (2\pi)^{-\frac{1}{2}} \int_{-\infty}^{x} e^{-t^2/2}\, dt$,

(b) $P_A\{\sigma^{-1} z(1 + m + \ldots + m^{n-1})^{\frac{1}{2}}(\hat{m}-m) \leq x\} \to \int_0^\infty \Phi(x\sqrt{w})\, dS(w)$

and (c), the results (a) and (b) continue to hold if P_A is replaced by $P\{\cdot | Z_n > 0\}$.

Proof. We outline the main steps and refer to the quoted papers for details. By the central limit theorem the asymptotic distribution as $p \to \infty$ of

$$\sigma^{-1} p^{-\frac{1}{2}} \sum_{i=1}^{p} (Y_i - m)$$

is standard normal. Now

$$\sigma^{-1}(Z_0+\ldots+Z_{n-1})^{\frac{1}{2}}(\hat{m}-m) = \sigma^{-1}(Z_0+\ldots+Z_{n-1})^{-\frac{1}{2}} \sum_{i=1}^{Z_0+\ldots+Z_{n-1}} (Y_i - m)$$

and at least if $p_0 = P\{Z_1 = 0\} = 0$, the result (a) follows fairly directly from a central limit theorem for a sum of a random number of independent random variables such as quoted by Billingsley [16, Theorem 17.2]. Also, (b) is a consequence of that result and (c) is elementary since the sequence of sets $\{Z_n > 0\}$ decreases to A. The generalization to $p_0 > 0$ was carried through by Dion and Jagers in the above mentioned papers.

Notice that result (b) states that $\hat{m}$ is not asymptotically normal when deterministic normalizing constants are used. If the offspring distribution is geometric, then the limiting distribution of (b) is a Student-distribution with 2z degrees of freedom.

Remark. An iterated logarithm result for $\hat{m}$ was recently provided by Asmussen and Keiding [6].

Conditioning on the non-extinction set makes sense only in the supercritical case. However it would be interesting to study the asymptotic distribution of $\hat{m}$ (properly normalized) in the subcritical and critical cases, given $Z_n>0$. Results of this type were given by Pakes [63] for the estimators $\hat{p}_k$ discussed in Proposition 1.1 above.

We finally note that Yanev [72] provided asymptotic distribution theory under the assumption that both z and n become large. These results would presumably be quite important in practice. The character of the results turns out to depend on the usual criticality trichotomy.

Theorem 2.4 (Yanev [72]). Let $0<\sigma^2<\infty$. Then as $n\to\infty$ and $z\to\infty$

(a) $\hat{m} \overset{P}{\to} m$ and $E(\hat{m}) \to m$.

(b) If $m<1$, then the asymptotic distribution of

$$\left(\frac{z}{\sigma^2(1-m)}\right)^{\frac{1}{2}}(\hat{m}-m)$$

is standard normal.

(c) If m=1, then if $n/z\to 0$, the asymptotic distribution of

$$\frac{(zn)^{\frac{1}{2}}}{\sigma}(\hat{m}-m)$$

is standard normal, while if $n/z^2\to\infty$, the asymptotic distribution of

$$\frac{\sigma^2}{2z^2(1-\hat{m})}$$

is a stable distribution with exponent $\frac{1}{2}$.

(d) If $m>1$ and the offspring distribution has finite fourth moment, then the asymptotic distribution of

$$\sigma^{-1}\{z(1+\ldots+m^{n-1})\}^{\frac{1}{2}}(\hat{m}-m)$$

is standard normal.

Notice that the results in Theorems 2.1 (b) and 2.3 (b) correspond to (d) in the present theorem (since W will degenerate as $z\to\infty$.)

1.3 Alternative estimators of the offspring mean.

A different estimator of the offspring mean was proposed by Lotka [58] and studied by Nagaev [62], see also Crump and Howe [19] and Dion [22]. This estimator is defined by

$$\bar{m} = \begin{cases} Z_n/Z_{n-1} & \text{if } Z_{n-1} > 0 \\ 1 & \text{otherwise .} \end{cases}$$

Then as noted by Nagaev, $E(\bar{m} \mid Z_{n-1} > 0) = m$. Asymptotic properties as $Z_0 = z \to \infty$ are standard as before. If $1 < m < \infty$, it follows directly from Heyde's [33] results that as $n \to \infty$, $\bar{m} \to m$ a.s. on A. Crump and Howe [19] showed that if $0 < \sigma^2 < \infty$, $E(\bar{m} \mid A) \to m$ as $n \to \infty$.

Asymptotic distribution results as $n \to \infty$ were derived by Nagaev [62] and Dion [22]. In the supercritical case $m > 1$ these state that

$$P_A\{\sigma^{-1} Z_{n-1}^{\frac{1}{2}} (\bar{m}-m) \le x\} \to \Phi(x),$$

$$P_A\{\sigma^{-1} m^{n/2} (\bar{m}-m) \le x\} \to \int_0^\infty \Phi(x\sqrt{w})\, dS(w)$$

and that these limiting distribution results continue to hold if conditioning on A is replaced by conditioning on $\{Z_{n-1} > 0\}$.

Nagaev also provided results for the critical and subcritical cases; iterated logarithm laws may be obtained as before.

Finally Heyde [35] mentioned the possibility of using $m^* = Z_n^{1/n}$ as an estimator for m. If $E(Z_1 \log Z_1) < \infty$, $Z_n m^{-n} \to W$ a.s. as $n \to \infty$, and therefore $m^* \to m$ a.s. on A. The rate of convergence is illustrated by the fact that by an application of Heyde and Leslie [37, Theorem 2], one has $n(m^* - m) \to m \log W$ a.s. on A. This "linear" rate of convergence compares badly to the "geometric" rate of $\hat{m}$ and $\bar{m}$, as pointed out by Heyde.

1.4 Estimation of the offspring variance.

In order to use results from Sections 2 and 3 to deduce confidence intervals for the mean of the offspring distribution, one has to give a consistent estimator for the variance σ^2. Such an estimator is given by

$$(4.1) \quad \overset{*}{\sigma}{}^2 = n^{-1} \sum_{k=0}^{n-1} Z_k((Z_{k+1}/Z_k - m^*)^2)$$

where m^* is either Harris' estimator $\hat{m}$ (Dion [23]) or the Lotka-Nagaev estimator $\bar{m}$ (Heyde [34]). Denote $\overset{*}{\sigma}{}^2$ by $\hat{\sigma}^2$ if $m^*=\hat{m}$ and by $\bar{\sigma}^2$ if $m^*=\bar{m}$. The properties of $\overset{*}{\sigma}{}^2$ are best studied by taking first $m^*=m$ and then replacing m by either $\hat{m}$ or $\bar{m}$.

The asymptotic behaviour of $\overset{*}{\sigma}{}^2$ has been studied in the case when $p_0=0$, but one expects the same asymptotic results to hold on the set of non-extinction in the general situation where $p_0>0$ is allowed.

Theorem 4.1 (Dion [23]) Assume $p_0 = 0$, $m>1$, $0<\sigma^2<\infty$ and $E(Z_1{}^4)<\infty$. Put $\tilde{\sigma}^2=n^{-1} \sum_{k=0}^{n-1} Z_k((Z_{k+1}/Z_k - m)^2)$. Then

(a) $(\tilde{\sigma}^2-\hat{\sigma}^2)n^{1-\varepsilon} \overset{P}{\to} 0, \quad \forall\varepsilon>0$.

(b) $\tilde{\sigma}^2 > \hat{\sigma}^2$ a.s.

Proof. For all $\varepsilon>0$,

$$(\tilde{\sigma}^2-\hat{\sigma}^2)n^{1-\varepsilon} = n^{-\varepsilon} \sum_{k=0}^{n-1} Z_k[(\hat{m}-m)\{2(Z_{k+1}/Z_k)-(\hat{m}+m)\}]$$

$$= (Z_0+\ldots+Z_{n-1})(\hat{m}-m)^2 . n^{-\varepsilon} > 0 \text{ a.s.},$$

from which (b) follows. By Theorem 2.3(a) the asymptotic distribution of $\sigma^{-2}(Z_0+\ldots+Z_{n-1})(\hat{m}-m)^2$ is χ^2 with 1 d.f., and therefore

$$|\tilde{\sigma}^2-\hat{\sigma}^2| n^{1-\varepsilon} \overset{P}{\to} 0, \quad \forall\varepsilon>0.$$

Note 1. It is easy to show that $E(\tilde{\sigma}^2)=\sigma^2$. The property (b) then states that $E(\hat{\sigma}^2)<\sigma^2$.

Note 2. By straightforward but long computations with conditional expectations and variances one can show that $\text{Var}(\tilde{\sigma}^2) \approx 2\sigma^4/n$. From property (a), $\hat{\sigma}^2$ is then a consistent estimator of σ^2.

Note 3. Using a central limit theorem for martingales due to Billingsley [15, p.52], the asymptotic normality of $\tilde{\sigma}^2$ (and hence of $\hat{\sigma}^2$) can be easily established under the further assumption $E(Z_1^6) < \infty$. But as is seen in the next theorem Heyde [34] has stronger results for $\bar{\sigma}^2$. These have been extended to $\hat{\sigma}^2$ by P. Feigin in an unpublished thesis.

Theorem 4.2 (Heyde [34]). Assume $p_0 = 0$, $m > 1$ and $0 < \sigma^2 < \infty$. Then

(a) $\bar{\sigma}^2 \overset{a.s.}{\to} \sigma^2$

(b) if $E(Z_1^4) < \infty$, then

$\sqrt{n}(\bar{\sigma}^2 - \sigma^2)/(2\sigma^4)^{\frac{1}{2}} \overset{D}{\to} N(0,1)$ and

(c) if $E(Z_1^{4+\delta})$ for some $\delta > 0$, then

$\bar{\sigma}^2 = \sigma^2 + \eta(n)(4\sigma^4 n^{-1} \log\log n)^{\frac{1}{2}}$,

where $\eta(n)$ has its set of a.s. limit points confined to $[-1,+1]$ with $\limsup_{n\to\infty} \eta(n) = +1$ a.s. and $\liminf_{n\to\infty} \eta(n) = -1$ a.s.

For the proof, which is rather long and far from trivial, the reader is referred to Heyde's article. However we will give an idea of his proof which relies heavily on strong results for martingales.

Let F_n be the σ-field generated by $Z_1, \ldots, Z_n$ and put

$$U_{k+1} = (Z_{k+1} - mZ_k)^2 Z_k^{-1} - \sigma^2.$$

Using once more the underlying independent r.v. that constitute the family tree, one has that the U_k are martingale differences. Apply next a strong law of large numbers for martingales to conclude that

$$n^{-1} \sum_{k=1}^{n} U_k \overset{a.s.}{\to} 0$$

and then the property (a) follows after proving that m can be replaced by $\overline{m}$. (This last fact still requires an iterated logarithm analogue for $\overline{m}$ as given by Heyde & Leslie [37]).

The central limit result (b) is obtained from a central limit theorem for martingales (Theorem 2 of Brown [17]) applied to the martingale $\{\sum_{j=1}^{n} U_j, F_n\}$ and the iterated logarithm result (c) is deduced from a corresponding iterated logarithm result for martingales due to Heyde and Scott [38, Theorem 1].

In the critical case, $m=1$, Nagaev [62] observed that $\beta_n = (Z_n - Z_{n-1})^2/Z_{n-1}$ could be used to estimate σ^2 since $E(\beta_n \mid Z_{n-1} > 0) = \sigma^2$ and if $E(Z_1^4) < \infty$, Var $(\beta_n \mid Z_{n-1} > 0) = O(\log n/n)$.

1.5 Immigration.

Consider now $X_0=1, X_1, X_2, \ldots$ a Galton-Watson process with immigration whose offspring distribution has the distribution of Z_1 with $EZ_1 = m$ and $0 < \mathrm{Var}(Z_1) = \sigma^2 < \infty$. Suppose further that the immigration process has a finite mean $\lambda = \Sigma\, ib_i$.

Estimation problems in that context have been studied by Heyde & Seneta [39,40,41] and Quine [65].

In the supercritical case, m>1, the asymptotic properties of the Lotka-Nagaev estimator as given in Section 3 and Heyde's [34] estimator for σ^2 are both robust in the sense that the asymptotic results as $n \to \infty$ continue to apply unchanged in the case where immigration occurs. Only minor modifications of the proofs for the Galton-Watson process without immigration are necessary to establish these results.

The subcritical case (m<1) however presents a new situation. Put $\lambda_1=m$ and $\lambda_2=\lambda$. The estimators investigated are

$$\hat{\lambda}_1 = \frac{\sum_{i=1}^{n} X_i (X_{i+1} - n^{-1} S_n)}{\sum_{i=1}^{n} (X_i - n^{-1} S_n)^2}$$

and

$$\hat{\lambda}_2 = \frac{S_n}{2n} \frac{\sum_{i=1}^{n} (X_{i+1}-X_i)}{\sum_{i=1}^{n} (X_i-n^{-1}S_n)^2}$$

where $S_n = \sum_{i=1}^{n} X_i$. Let $\mu = \lambda_2(1-\lambda_1)$, $c^2 = \mu\sigma^2 + b^2$ where b^2 is the variance of the immigration distribution. Assume further the immigration distribution is not degenerate. Then, improving on the work by Heyde & Seneta [40,41], Quine [65] proved that $\hat{\lambda}_1$ and $\hat{\lambda}_2$ are strongly consistent and obey the central limit theorem and law of the iterated logarithm under the sole condition $c^2 < \infty$.

Theorem 5.1 (Quine [65]). If $c^2 < \infty$, then for $i = 1,2$ as $n \to \infty$

(a) $\hat{\lambda}_i \to \lambda_i$ almost surely.

(b) The asymptotic distribution of $\sqrt{n}(\hat{\lambda}_i - \lambda_i)/k_i$ is standard normal, and

(c) $$\limsup \frac{n^{\frac{1}{2}}(\hat{\lambda}_i - \lambda_i)}{k_i(2 \log \log n)^{\frac{1}{2}}} = 1 \quad \text{a.s.}$$

$$\liminf \frac{n^{\frac{1}{2}}(\hat{\lambda}_i - \lambda_i)}{k_i(2 \log \log n)^{\frac{1}{2}}} = -1 \quad \text{a.s.}$$

as long as $k_i < \infty$, where k_i are constants given by Quine [65, p.320, cf. the correction note].

The theorem follows from known results about Markov chains. To illustrate that, let us recall the very elegant proof of (a) as given by Quine. Note that the state space I contains a countable irreducible set I* on which $X_1, X_2, \ldots$ have their support. Clearly I* is aperiodic since

$$\kappa \equiv \inf\{i : b_i > 0\} = \inf\{i : i \in I^*\}$$

is accessible at all times $n \geqq 1$. Furthermore as $\Sigma b_j \log j < \infty$, I* forms a positive class. Let $\overline{T}_{I*}$ be the σ-field of all subsets of $I*$. If φ is any function from $F_{I*} \times F_{I*}$ to R, and if $\{\pi_i\}$ is the limiting distribution of $\{X_n\}$, it follows from Billingsley [15, Theorems 1.1 and 1.3] that

$$n^{-1}\Sigma\, \varphi(X_j, X_{j+1}) \to \sum_{i,j} \pi_i P(X_1=j \mid X_0=i)\varphi(i,j) \quad \text{a.s.}$$

as long as the limiting series converges absolutely. It can be shown that the stationary distribution $\{\pi_i\}$ has mean μ and variance $c^2/(1-\lambda_1^2)$.

There remains only to choose suitable φ to show that

$$n^{-1} S_n \to \mu \quad \text{a.s.}$$

$$n^{-1} \sum_{i=1}^{n} (X_{i+1} - X_i)^2 \to 2c^2/(1+\lambda_1) \quad \text{a.s.}$$

$$n^{-1} \sum_{i=1}^{n} X_i^{\,2} \to c^2/(1-\lambda_1^{\,2}) + \mu^2 \quad \text{a.s.}$$

and

$$n^{-1} \sum_{i=1}^{n} (X_i - \mu)^3 \to \gamma \quad \text{a.s.}$$

The property (a) follows from these a.s. convergence results.

We conclude this section by calling attention to the interesting historical remarks and discussions of applications provided by Heyde & Seneta [40].

1.6 Multitype processes

This section contains a brief survey of some very recent results concerning the estimation of the mean matrix M, in particular the growth rate ρ in the positive regular case where ρ is given as the unique largest positive eigenvalue of M.

Motivated by a wish to be able to estimate at an early stage whether an epidemic is minor or major Becker [13] posed the problem of the behaviour of the three above mentioned estimators $\hat{m}$, $\overline{m}$ and $m*$ when the process is in fact a multitype Galton-Watson process

$(Z_n) = ((Z_n(1),\ldots,Z_n(p)))$ and the estimators are based on the total generation sizes $|Z_n| = Z_n(1)+\ldots+Z_n(p)$. Becker showed under mild conditions ($\rho > 1$ and "j log j") that all of these estimators will converge almost surely to ρ on the set of nonextinction as $n \to \infty$.

A more detailed analysis was provided by Asmussen and Keiding [6] who adapted martingale central limit theory to martingale difference triangular arrays indexed by the set of all individuals ever alive. Aside from $(|Z_1|+\ldots+|Z_n|)/(|Z_0|+\ldots+|Z_{n-1}|)$, which we shall henceforth denote by $\tilde{\rho}$, these authors concentrated their interest on the obvious estimator $\hat{M} = (\hat{m}_{ij})$ of the whole mean matrix based on the observation of all combinations of types of parents and offspring. Let U^{ν}_{ki} be the vector of offspring of the k'th individual of type i in generation ν, $\nu = 0,\ldots,n-1$. Then

$$\hat{m}_{ij} = \sum_{\nu=0}^{n-1} Z^i_{\nu+1}(j) / \sum_{\nu=0}^{n-1} Z_\nu(i)$$

where $Z^i_{\nu+1}(j) = \sum_{k=1}^{Z_\nu(i)} U^{\nu}_{ki}(j)$ is the number of individuals of type j in the $(\nu+1)$st generation whose parents were of type i and $Z_\nu(j) = \Sigma Z^i_\nu(j)$. As noted by Keiding and Lauritzen [54], $\hat{M}$ is the maximum likelihood estimator of M whether based on observations on the U^{ν}_{ki} or on the $Z^i_\nu(j)$ only, in the completely general model as well as in models specified by classes of offspring distributions as discussed in Theorem 1.3 above.

In the results below, we assume throughout that M is positive regular, its principal eigenvalue $\rho > 1$, and zero offspring always has probability zero, so that $W = \lim \text{ a.s. } \rho^{-n} Z_n > 0$ a.s. Finally v and u are the left and right eigenvectors, $vM = \rho v$ and $Mu = \rho u$.

<u>Theorem 6.1</u> As $n \to \infty$,

(a) $\hat{M} \to M$ a.s.

(b) The limiting distribution of the matrix $([W(1+\rho+\ldots+\rho^{n-1})v(i)]^{\frac{1}{2}} (\hat{m}_{ij} - m_{ij}))$

is that of $(Y_i(j))$, where $Y_1,\ldots,Y_p$ are independent and the distribution of Y_i is p-dimensional normal with mean zero and variance matrix Σ^i given by

$$\Sigma^i_{jj} = \mathrm{Var}(Z_1(j) \mid Z_0 = e_i),$$

$$\Sigma^i_{jk} = \mathrm{Cov}(Z_1(j), Z_1(k) \mid Z_0 = e_i).$$

By the transformation invariance of maximum likelihood estimators, the maximum likelihood estimator $\hat{\rho}$ of ρ is given as the largest positive eigenvalue of $\hat{M}$, which will be well-defined at least for large n. Standard techniques then immediately yield

Corollary. As $n \to \infty$,

(a) $\hat{\rho} \to \rho$ a.s.

(b) The asymptotic distribution of $[W(1+\rho+\ldots+\rho^{n-1})]^{\frac{1}{2}}(\hat{\rho}-\rho)$ is normal $(0, v \cdot \mathrm{Var}^{\bullet} Z_1 \cdot u)$.

Iterated logarithm laws were also derived by Asmussen and Keiding.

We next turn our attention to the estimator $\tilde{\rho} = (|Z_1|+\ldots+|Z_n|) \,/\, (|Z_0|+\ldots+|Z_{n-1}|)$. Becker [13] proved strong consistency. The rate of convergence turns out to depend in a qualitative way on the rate of convergence of the relative type distribution to the stable type distribution, cf. Kesten and Stigum [57] and Asmussen [5]. As it appears from the latter references, there is a trichotomy depending on the relative sizes of $|\lambda|^2$ and ρ, where λ is a certain other eigenvalue of M. In fact, if $|\lambda|^2 \geqq \rho$, no central limit result parallel to that of the Corollary above holds.

Theorem 6.2 When $|\lambda|^2 < \rho$, as $n \to \infty$, $[W(1+\rho+\ldots+\rho^{n-1})]^{\frac{1}{2}}(\tilde{\rho}-\rho)$ is asymptotically normal with mean zero and variance computed by Asmussen and Keiding [6].

We finally note that Quine and Durham [66] recently obtained multitype analogues of the results by Heyde and Seneta [40,41] and Quine [65] concerning the subcritical process with immigration and discussed in Section 5 above.

1.7 Random environments

Becker [13] continued his investigation by inquiring into the properties of $\hat{m}$, $\bar{m}$ and m^* when the process (Z_n) is in fact a branching process with (independent identically distributed) random environments. As is well known, it is here important to distinguish between the average offspring mean $\mu = E(Z_1 \mid Z_0 = 1)$ and the growth rate (criticality parameter) $\tau = \exp(E[\log Z_1 \mid Z_0, \zeta_1])$, ζ_1 denoting the environment of the first generation. To Becker, τ was of prime importance and he observed that m^* is strongly consistent for τ whereas

$$\tilde{m} = n^{-1} \sum_{j=1}^{n} (Z_j/Z_{j-1})$$

is strongly consistent for μ. An independent study by Dion and Esty [24] also pointed out these facts and went on to derive asymptotic variances and asymptotic distribution results for m^* and $\tilde{m}$.

1.8 Estimation of the age of the process

It has been suggested that it might occasionally be of interest to estimate the generation number of a Galton-Watson process, assuming that it descended from one ancestor, see Stigler [69] and Crump and Howe [19]. For a discussion of the possibility of applying this idea to rare human blood types we refer to Thompson [70,71]. It appears from Thompson's analysis that the evaluation of the age is critically dependent on the estimated mean offspring, and at least in her context it seemed difficult to estimate both at the same time.

2. ESTIMATION THEORY FOR CONTINUOUS-TIME BRANCHING PROCESSES

2.1 Maximum likelihood estimation of parameters in Markov branching processes

Consider a Markov branching process $(X_t, t \geq 0)$ with split intensity $\lambda > 0$ and offspring distribution $(p_k, k = 0,1,2,\ldots, \Sigma p_k = 1)$. As usual we assume $p_1 = 0$ and $X_0 = x_0$ degenerate.

If this process is assumed observed continuously over a time interval $[0,t]$, and if it is known that with probability one only finitely many jumps happen in each finite interval (a sufficient condition being that the offspring mean $m = \Sigma kp_k < \infty$, cf. Harris [32, p.107]), then it follows by a minor modification of Albert's [4] arguments that a measure may be constructed such that the likelihood function is as given in the Theorem below.

Theorem 2.1 The likelihood function corresponding to observation of $\{X_u : 0 \leq u \leq t\}$ is

$$\lambda^{N_t} e^{-\lambda S_t} \prod_{k=0,2}^{\infty} p_k^{N_t(k)}$$

where $N_t(k)$ is the number of splits of size k-1 in $[0,t]$, $N_t = \Sigma N_t(k)$ is the total number of splits, and

$$S_t = \int_0^t X_u \, du$$

is the total time lived by the population (the total exposure time) in $[0,t]$.

For the statistical problem specified by $\lambda \in (0,\infty)$ and all offspring distributions with finite mean the maximum likelihood estimators are given as

$$\hat{\lambda} = N_t/S_t, \quad \hat{p}_k = N_t(k)/N_t$$

when $N_t > 0$.

Remark. The results of the Theorem continue to hold with the obvious changes if the deterministic observation interval [0,t] is replaced by [0,T], where T is a stopping time not depending on the parameters. If in particular T is chosen as a first hitting time for N_t or S_t, the distribution of $\hat{\lambda}$ will be more easily tractable, since it is no longer a ratio between random variables, cf. Moran [60,61] and Kendall [55].

Corollary. The maximum likelihood estimator of the offspring mean m and the growth rate (Malthusian parameter) $\alpha = (m-1)\lambda$ are given by

$$\hat{m} = 1 + (X_t - x_0)/N_t, \quad \hat{\alpha} = (X_t - x_0)/S_t.$$

A further look at the estimation problem reveals that as long as the split parameter λ and the parameters of the offspring distribution are variation independent, there are two separate problems: The random number N_t of i i d observations of the offspring distribution, leading to standard relative frequency estimators, and the question of estimating the split intensity λ, where the occurrence/exposure rate N_t/S_t is seen to arise in a natural way. As will be seen below, the latter question is of greatest theoretical interest and will be emphasized.

It follows from these remarks that no great new theoretical insight is to be expected from the analysis of more restricted models based on particular classes of offspring distributions. Aside from the general model discussed above we shall therefore only mention the linear birth-and-death process with $p_0+p_2 = 1$, in particular the pure birth (Yule) process ($p_2 = 1$) and the pure death process ($p_0 = 1$).

Very few small-sample results exist. Beyer, Keiding and Simonsen [14] gave formulas, numerical tables, and approximative results for the first three moments of the maximum likelihood estimator in the pure birth process and the pure death process. That paper also surveyed earlier literature.

As in Chapter 1, we shall therefore have to be satisfied mostly with asymptotic results.

2.2 Asymptotic results for estimators in Markov branching processes.

As for the time-discrete case, two types of asymptotic theory are relevant: large number of ancestors and large interval of observation.

Though the many-ancestors theory might often be the more important in practice, it is a fairly standard application of i i d estimation theory and we shall therefore be content with referring to the statement of the results by Keiding [49,51] and Athreya and Keiding [7].

The results for $t \to \infty$ are of greatest interest in the super-critical case $\alpha > 0$ ($\Leftrightarrow m > 1$) since otherwise $X_t \to 0$ a.s. Let $A = \{X_t \to 0\}$.

Theorem 2.1 As $t \to \infty$,

(a) if $m < \infty$, then $\hat{\lambda} \to \lambda$ and $\hat{p}_k \to p_k$, $k = 0,2,3,\ldots$, a.s. on A.

(b) if $0 < \sigma^2 < \infty$ (where σ^2 is the offspring variance $\Sigma(k-m)^2 p_k$), then the asymptotic distribution, given A, of

$$S_t^{\frac{1}{2}}(N_t(0)/S_t - \lambda p_0,\ N_t(2)/S_t - \lambda p_2,\ N_t(3)/S_t - \lambda p_3, \ldots)$$

is that of independent normals with parameters $(0, \lambda p_k)$, $k = 0,2,3,\ldots$

Remark. Notice that $N_t(k)/S_t = \hat{\lambda}\hat{p}_k$ is the maximum likelihood estimator of λp_k.

Corollary. As $t \to \infty$,

(a) if $m < \infty$, $\hat{\alpha} = (X_t - x_0)/S_t \to \alpha$ a.s. on A,

(b) if $\sigma^2 < \infty$, the asymptotic distribution, given A, of $S_t^{\frac{1}{2}}(\hat{\alpha} - \alpha)$, is normal $(0, \lambda(\sigma^2 + (m-1)^2))$.

Remark. These statements may be transformed into statements with deterministic normalising factors in a similar way as discussed for the time-discrete case in Section 1.2 above. Also, for the purpose of suggesting asymptotic confidence intervals, conditioning on the set A may be replaced by conditioning on $\{X_t > 0\}$.

As indicated in Section 1 above, the estimators of the offspring probabilities are really just the standard relative frequencies, although with a random number of replications. To give an indication of methods of proof we may therefore restrict ourselves to the pure birth process with $p_2 = 1$.

Thus let $(X_t, t \geqq 0)$ be a pure birth process with birth intensity λ and $X_0 = x$ degenerate. We want to prove, with $\hat{\lambda} = (X_t - x)/S_t$, $S_t = \int_0^t X_u \, du$,

Theorem 2.2 As $t \to \infty$,

(a) $\hat{\lambda} \to \lambda$ a.s.,

(b) the asymptotic distribution of $(\lambda S_t)^{\frac{1}{2}}(\hat{\lambda}/\lambda - 1)$ is standard normal.

We shall indicate three different methods of proof, the two first of which are applicable for general Markov branching processes.

First proof. (Keiding [51]). Define the stochastic process V_u by $S_{V_u} = u$ (for each ω, $V.(\omega)$ is the inverse function of the continuous sample function $S.(\omega)$). Then it is well-known (Athreya and Ney [8, Theorem III.11.1], Papangelou [64], Rudemo [68] that

$$Y_u = X_{V_u} - x, \quad u \geqq 0$$

is a homogeneous Poisson process with intensity λ. This having i i d increments, it follows directly from the strong law of large numbers that $Y_u/u \to \lambda$ a.s. as $n \to \infty$. But hence

$$(X_t - x)/S_t = Y_{S_t}/S_t \to \lambda$$

a.s. on $\{S_t \to \infty\}$, that is, a.s.

Moreover, by the central limit theorem $(\lambda u)^{-\frac{1}{2}}(Y_u - \lambda u)$ is asymptotically standard normal as $u \to \infty$. If we may substitute u by S_t, this proves part (b), since

$$(\lambda S_t)^{-\frac{1}{2}}(Y_{S_t} - \lambda S_t) = (\lambda S_t)^{\frac{1}{2}}(\hat{\lambda}/\lambda - 1).$$

It follows from central limit theory for sums of a random number of i i d random variables that this substitution is permissible when as here $S_t e^{-\lambda t} \to W/\lambda$ a.s., where $P\{0 < W < \infty\} = 1$, cf. e.g. Billingsley [16, Section 17].

<u>Second proof</u>. (Athreya and Keiding [7]). Let $0 = T_0 < T_1 < T_2 < \dots$ be the birth times of (X_t), that is,

$$X_{T_j} = x + j, \quad X_{T_j -} = x + j - 1.$$

Let $Z_i = X_{T_{i-1}}(T_i - T_{i-1})$, $Z'_t = X_{t-}(t - T_{X_t - x})$. It is then well known, cf. Athreya and Ney [8, p.127] that

$$S_t = \sum_{i=1}^{X_t - x} Z_i + Z'_t$$

where $Z_1, Z_2, \dots$ are i i d exponential with intensity λ. Therefore $(Z_1 + \dots + Z_n)/n \to \lambda^{-1}$ a.s. and hence (granted the easily shown fact that $Z'_t/X_t \to 0$ a.s.)

$$\hat{\lambda}^{-1} = S_t/(X_t - x) \to \lambda^{-1} \text{ a.s. on } \{X_t \to \infty\}.$$

Also, the asymptotic distribution of

$$\lambda n^{-\frac{1}{2}}\left(\sum_{i=1}^{n} Z_i - n\lambda^{-1}\right)$$

is standard normal, and by again using random sum central limit theory it may be concluded that

$$S_t(X_t - x)^{-\frac{1}{2}}(\lambda - \hat{\lambda}) = \lambda(X_t - x)^{-\frac{1}{2}}\left(\sum_{i=1}^{X_t - x} Z_i + Z'_t - (X_t - x)\right)$$

is asymptotically standard normal, from which part (b) follows by noticing that $[S_t/(X_t - x)]^{\frac{1}{2}} \overset{P}{\to} \lambda^{-\frac{1}{2}}$.

Third proof. (Keiding [49]. An approach which does not seem amenable to generalization beyond the pure birth process is the following. Let as usual $W = \lim$ a.s. $X_t e^{-\lambda t}$. Then the stochastic process $X_t - x$ has, conditional on $W = w$, the same distribution as an (inhomogeneous) Poisson process with intensity function $x w \lambda e^{\lambda t}$ (Kendall [56]).

The strong consistency and asymptotic normality may now be proved by direct calculation in the Poisson process. - Besides yielding these asymptotic results, this latter approach might suggest possible conditional inference procedures. We refer to Keiding [49] for further discussion.

2.3 Estimation in multitype Markov branching processes.

The simple estimators studied so far for the single-type processes generalize in an obvious way to multitype processes as long as all individual offspring may be observed, which it seems natural to assume in a genuine continuous time context. Details were given by Athreya and Keiding [7] and will not be reproduced here.

2.4 Estimation theory for more general continuous-time branching processes.

The simple occurrence/exposure rate $(X_t - x)/S_t$ is obviously strongly consistent for the Malthusian parameter α in any branching process. To elucidate its rate of convergence, Athreya and Keiding [7] remarked that any Bellman-Harris process will for large t in a certain sense look very much like a Markov branching process. Asymptotic normality was conjectured but later disproved by Asmussen and Keiding [6] who showed that in fact the rate of convergence of the age distribution becomes essential, much in the same way as was the case in the discrete-time, multitype case, cf. Section 1.6 above.

Estimation problems in cell kinetics based on the framework of the general branching process defined by Crump and Mode [20] and Jagers [43] have been studied by Jagers [45] and Jagers and Norrby [47]. Hoel and Crump [42] studied alternative estimators of the parameters of the generation time distribution in a Bellman-Harris process.

It should finally be mentioned that Brown and Hewitt [18] studied inference for the diffusion branching process.

3. THE ROLE OF MARTINGALE THEORY; EFFICIENCY CONCEPTS

3.1 The score function as a martingale

Let $X_1,\ldots,X_p$ be i i d random variables from a one-dimensional exponential family with density $a(\theta)b(x)e^{\theta t(x)}$, $\theta \in \Theta$, and let $\tau(\theta) = E_\theta[t(X_i)]$. Then under suitable regularity conditions the statistical model may be parameterized by τ and the score function based on the p replications is

$$S_p(\tau) = D_\tau \log L_p = p\, i(\tau)(\overline{T} - \tau)$$

with L_p the likelihood function, $\overline{T} = \Sigma t(X_i)/p$ and $i(\tau) = -D_\tau^2 \log L_p$ the usual Fisher information in one observation. (See Barndorff-Nielsen [9] for exponential family theory). Notice that the maximum likelihood estimator $\hat{\tau}$ of τ is exactly $\overline{T}$, and that

$$\mathrm{Var}\,(S_p(\tau)) = p\, i(\tau) = [\mathrm{Var}\,(\hat{\tau})]^{-1}$$

so that each new observation adds an amount of $i(\tau)$ to the information content or "precision" of the experiment. For purposes of later generalization we also remark that the stochastic process $(S_p(\tau),\ p = 1,2,\ldots)$ has i i d increments with mean zero and finite variance, and hence is a square integrable (local) martingale. The Doob decomposition of $S_p^2(\tau)$ is $S_p^2(\tau) = I_p(\tau) + \text{martingale}$, where

$$I_p(\tau) = \sum_{j=0}^{p-1} \mathrm{Var}\,(S_{j+1}(\tau) \mid X_1,\ldots,X_j) = p\, i(\tau).$$

In this sense the increasing process corresponding to $S_p^2(\tau)$ exactly measures the information content.

For generalizations of these ideas to the branching process context it is useful to concentrate on linear discrete one-parameter exponential families, also called power series families, where the canonical statistic $t(X) = X$. As just stated, such a family may be parameterized by the mean m. It now follows from the discussion in Section 1.1 that if the offspring distribution is assumed to belong to a power series family, then the score function $S_n(m)$ corresponding to observation of the first $n+1$ generation sizes $Z_0,\dots,Z_n$ of a Galton-Watson process is obtained by replacing p by $Z_0+\dots+Z_{n-1}$, so that

$$S_n(m) = (Z_0+\dots+Z_{n-1})\, i(m)(\hat{m} - m),$$

where, as in Chapter 1, $\hat{m} = (Z_1+\dots+Z_n)/(Z_0+\dots+Z_{n-1})$.

It is quite generally true and easily directly verified in this particular case that $(S_n(m),\ n = 0,1,\dots)$ is a square integrable (local) martingale with respect to the "self-exciting" family of σ-algebras $(\mathcal{N}_n = \sigma\{Z_0,\dots,Z_n\},\ n = 0,1,2,\dots)$.

The Doob decomposition of the submartingale $(S_n^2(m), \mathcal{N}_n)$ is given by

$$S_n^2(m) = I_n(m) + \text{martingale},$$

where

$$I_n(m) = \sum_{j=0}^{n-1} [E(S_{j+1}^2(m) \mid \mathcal{N}_j) - S_j^2(m)]$$

is increasing. We may alternatively write

$$I_n(m) = \sum_{j=0}^{n-1} E[(S_{j+1}(m) - S_j(m))^2 \mid \mathcal{N}_j]$$

where the individual term in our case is

$$E[\{i(m)(Z_{j+1} - mZ_j)\}^2 \mid \mathcal{N}_j] = i(m)^2 \operatorname{Var}(Z_{j+1} \mid \mathcal{N}_j) = Z_j\, i(m)$$

so that we have

$$(1.1) \quad I_n(m) = (Z_0+\dots+Z_{n-1})\, i(m)$$

and the score function may be written

(1.2) $S_n(m) = I_n(m)(\hat{m} - m)$.

This very simple form led Heyde [35] and Heyde and Feigin [36] to several definitions and generalizations, interpreting quite generally the increasing process $I_n(m)$ as a measure of information content. The form

$$I_n(\tau) = \sum_{j=0}^{n-1} \text{Var}\,(S_{j+1}(\tau) \mid N_j)$$

illustrates that I_n successively accumulates whatever new information there is in taking an extra observation, given the previous samples.

When as in the present branching process context (cf. (1.1)) the information function $I_n(m)$ factorises into a product of a deterministic factor and an observable quantity, Heyde and Feigin termed the statistical model a "conditional exponential family". We notice that this concept generalizes the role of mean value parameterized exponential families in the i i d theory.

These ideas carry over to continuous time, with the Doob-Meyer decomposition of the squared score function as a key tool, see Feigin [27]. For the Yule process studied in Section 2.2 above, the information content becomes $I_t(\lambda) = S_t/\lambda$ with $S_t = \int_0^t X_u\,du$. A general framework for statistical models for point processes whose conditional intensities factorize into a product of a deterministic function and an observable process was given by Aalen [1,2,3] who also leaned heavily on martingale theory.

Turning next to asymptotic theory, we demonstrated in the previous chapters how it is possible to reduce all asymptotic results for one-type processes to the basic i i d structure inherent in branching processes. The analysis by Asmussen and Keiding [6] indicated, however, that already for the multitype Galton-Watson process, this may no longer be feasible. At least as a starting point for a methodology for more general stochastic process estimation problems, it is useful to point out that a martingale central limit theorem with random norming factors will yield asymptotic normality of

$$I_n^{-\frac{1}{2}}(m)S_n(m) = I_n^{\frac{1}{2}}(m)(\hat{m}-m)$$

which is equivalent to Theorem 1.2.3 (a) for the particular processes here discussed. Of course consistency is even more directly shown. The martingale approach is based only on the rather generally valid martingale property of the score function and will thus have considerably wider applicability.

3.2 Efficiency concepts.

Heyde [35] proposed to call a consistent estimator T_n of θ asymptotically efficient if there exists a (deterministic) function $\beta(\theta)$ such that

$$I_n^{\frac{1}{2}}(\theta)[T_n - \theta - \beta(\theta)S_n(\theta)/I_n(\theta)] \overset{P}{\to} 0$$

as $n \to \infty$. This definition reduces to that of Rao [67, Sec. 5c] in the i i d case. Heyde proved that the maximum likelihood estimator of the offspring mean in power series families (including the geometric distributions discussed above) is asymptotically efficient in this sense.

A complication with the use of the maximum likelihood estimator here is that it is not sufficient. Indeed, from (1.2) it is seen that since $I_n(m)$ is random, the two-dimensional statistic $(\hat{m}, Z_0 + \ldots + Z_{n-1})$ is minimal sufficient. Generalizing an argument due to Rao on the local behaviour of the power function near the null hypothesis Basawa and Scott [10] pointed out that asymptotically, hypothesis testing based on the maximum likelihood estimator alone is inferior to tests based on the full sufficient statistic. A different analysis by Feigin [29] based on the concept of contiguity raised some doubts about Basawa and Scott's approach.

A general framework for statistical models with one "curved" (nonlinear) submodels of exponential families was provided by Efron [25] who coined the term "statistical curvature" for a quantity entering into the asymptotic distributions of estimators and test statistics. In the discussion of Efron's paper, Keiding [52] mentioned the possible role of this concept in branching process situations, and further work along these lines, so far unpublished, has been done by I.V. Basawa.

4. APPLICATIONS

The area of statistical inference for branching processes is closely related to classical life testing, certain problems in reliability, and more general statistical problems for point processes, all of which are very important for the applications. However, it is not yet clear how far branching process models such as those discussed in this paper will be directly applicable as statistical models. Although a branching process may be a useful description of the underlying stochastic phenomena, the important statistical problems might often derive from incomplete observation, random temporal variation of parameters, or plain measurement uncertainty, and none of these sources of random variation are included in the present theory (save for the preliminary results concerning BPRE and quoted in Section 1.7). Furthermore, the number of replications will presumably often be so large that the intrinsic random variation discussed in this paper will be negligible.

We have already mentioned some specific applications of methods like the ones discussed here: the estimation of the age of rare human blood types (Section 1.8) and various questions in cell kinetics (Section 2.4). For an example where a branching process model leads to quite classical statistical problems, see Gani and Saunders [30].

We conclude the paper by quoting two further applications of the present theory that we have come across.

4.1 Population dynamics of the whooping crane population of North America

The whooping crane is an extremely rare migratory bird with breeding area in western Canada. Miller et al. [59] reported the annual counts from 1938-1972 of whooping cranes arriving to the wintering grounds in Texas. The birds born the previous spring have a different plumage (they are referred to as "young") and it is therefore possible to obtain the annual number of births and deaths. The population has increased from 14 individuals in 1938 to 57 in 1970. Miller et al. fitted a simple linear birth-death process and used the estimated values to produce prediction intervals concerning future population sizes. A critical discussion and suggestions of other possible models were provided by Keiding [53] and a time series approach is due to Kashyap and Rao [48, p. 296].

4.2 Epidemics: early evaluation of whether an epidemic is minor or major.

In a series of papers Becker [11,12,13] has developed and applied aspects of the theory of statistical inference for branching processes to estimate the initial infection rate of an epidemic. The method is to approximate the number of infectives in a stochastic epidemic process by a suitable Galton-Watson process, and then estimate the offspring mean of the latter. It is also possible to study the impact of vaccination by assuming that a fixed known proportion of the population in question is immunized.

Becker first [11] considered the application of the simple Galton-Watson process to the 49 smallpox epidemics in Europe between 1950 and 1970. These epidemics were all minor (became extinct) but the estimated mean offspring size was not significantly less than one which may raise the question whether major smallpox epidemics were only avoided by luck.

The approximation by a Galton-Watson process involves several assumptions. First, successive "generations" of infectives should be reasonably discrete, that is, there should be a fairly long latent period and a short infective period. Second, the number of susceptibles should be so large that it does not decrease appreciably when individuals get infected. As an example of a situation where this assumption clearly does not hold Becker [12] considered an observed (major) smallpox epidemic in a small, closed Nigerian community. To implement the theory for this small population, he approximated the number of infectives with a size-dependent Galton-Watson process and used a least squares approach to derive an estimator of the initial infection rate.

A third assumption is that of a homogeneously mixing population. It was mentioned in Sections 1.6 and 1.7 above that Becker [13] studied the fate of several estimators of the offspring mean on the assumption that the process is in fact multitype Galton-Watson or develops in random environments.

ACKNOWLEDGMENTS

We are grateful to Paul Feigin for critical comments and to several authors for providing us with prepublication versions of their work.

REFERENCES

[1] Aalen, O. O. Statistical inference for a family of counting processes. Ph. D. Dissertation, Department of Statistics, University of California, Berkeley, 1975.

[2] Aalen, O. O. Weak convergence of stochastic integrals related to counting processes. Z. Wahrscheinlichkeitsth. verw. Geb. 38, 261-277 (1977).

[3] Aalen, O. O. Nonparametric inference for a family of counting processes. Ann. Statist. (to appear 1978).

[4] Albert, A. Estimating the infinitesimal generator of a continuous time, finite state Markov process. Ann. Math. Statist. 33, 727-753 (1962).

[5] Asmussen, S. Almost sure behavior of linear functionals of supercritical branching processes. Trans. Amer. Math. Soc. 231, 233-248 (1977).

[6] Asmussen, S. & Keiding, N. Martingale central limit theorems and asymptotic estimation theory for multitype branching processes. Adv. Appl. Prob. 10 (to appear 1978).

[7] Athreya, K.B. & Keiding, N. Estimation theory for continuous-time branching processes. Sankhyā Ser. A (to appear 1976).

[8] Athreya, K.B. & Ney, P.E. Branching processes. Springer, Berlin, 1972.

[9] Barndorff-Nielsen, O. Information and exponential families in statistical theory. Wiley, London, 1978.

[10] Basawa, I.V. & Scott, D.J. Efficient tests for branching processes. Biometrika 63, 531-536 (1976).

[11] Becker, N. On parametric estimation for mortal branching processes. Biometrika 61, 393-399 (1974).

[12] Becker, N. Estimation for an epidemic model. Biometrics 32, 769-777 (1976).

[13] Becker, N. Estimation for discrete time branching processes with application to epidemics. Biometrics 33, 515-522 (1977).

[14] Beyer, J.E., Keiding, N. & Simonsen, W. The exact behaviour of the maximum likelihood estimator in the pure birth process and the pure death process. Scand. J. Statist. 3, 61-72 (1976).

[15] Billingsley, P. Statistical inference for Markov processes. University of Chicago Press, 1961.

[16] Billingsley, P. Convergence of probability measures. Wiley, New York, 1968.

[17] Brown, B.M. Martingale central limit theorems. Ann. Math. Statist. 42, 59-66 (1971).

[18] Brown, B.M. & Hewitt, J.I. Inference for the diffusion branching process. J. Appl. Prob. 12, 588-594 (1975).

[19] Crump, K.S. & Howe, R.B. Nonparametric estimation of the age of a Galton-Watson branching process. Biometrika 59, 533-538 (1972).

[20] Crump, K.S. & Mode, C.J. A general age-dependent branching process. J. Math. Anal. Appl. 24, 494-508 (1968).

[21] Dion, J.-P. Estimation des probabilités initiales et de la moyenne d'un processus de Galton-Watson. Thèse de doctorat, Université de Montreal, 1972.

[22] Dion, J.-P. Estimation of the mean and the initial probabilities of a branching process. J. Appl. Prob. 11, 687-694 (1974).

[23] Dion, J.-P. Estimation of the variance of a branching process. Ann. Statist. 3, 1183-1187 (1975).

[24] Dion, J.-P. & Esty, W.W. Estimation problems in branching processes with random environments. Manuscript, 1977.

[25] Efron, B. Defining the curvature of a statistical problem (with applications to second order efficiency). (with discussion). Ann. Statist. 3, 1189-1242 (1975).

[26] Eschenbach, W. & Winkler, W. Maximum-Likelihood-Schätzungen beim Verzweigungsprozess von Galton-Watson. Math. Operationsforsch. u. Statist. 6, 213-224 (1975).

[27] Feigin, P.D. Maximum likelihood estimation for continuous-time stochastic processes. Adv. Appl. Prob. 8, 712-736 (1976).

[28] Feigin, P.D. A note on maximum likelihood estimation for simple branching processes. Austr. J. Statist. (to appear 1977).

[29] Feigin, P.D. The efficiency criteria problem for stochastic processes. Stoch. Proc. Appl. (to appear 1977).

[30] Gani, J. & Saunders, I.W. Fitting a model to the growth of yeast colonies. Biometrics 33, 113-120 (1977).

[31] Harris, T.E. Branching processes. Ann. Math. Statist. 19, 474-494 (1948).

[32] Harris, T.E. The theory of branching processes. Springer, Berlin, 1963.

[33] Heyde, C.C. Extension of a result of Seneta for the supercritical Galton-Watson process. Ann. Math. Statist. 41, 739-742 (1970).

[34] Heyde, C.C. On estimating the variance of the offspring distribution in a simple branching process. Adv. Appl. Prob. 6, 421-433 (1974).

[35] Heyde, C.C. Remarks on efficiency in estimation for branching processes. Biometrika 62, 49-55 (1975).

[36] Heyde, C.C. & Feigin, P.D. On efficiency and exponential families in stochastic process estimation. In: Statistical distributions in scientific work (ed. G.P. Patil, S. Kotz, J.K. Ord) Reidel, Dordrecht Vol. 1, 227-240, 1975.

[37] Heyde, C.C. & Leslie, J.R. Improved classical limit analogues for Galton-Watson processes with or without immigration. Bull. Austr. Math. Soc. 5, 145-156 (1971).

[38] Heyde, C.C. & Scott, D.J. Invariance principles for the law of the iterated logarithm for martingales and processes with stationary increments. Ann. Prob. 1, 428-436 (1973).

[39] Heyde, C.C. & Seneta, E. Analogues of classical limit theorems for the supercritical Galton-Watson process with immigration. Math. Biosc. 11, 249-259 (1971).

[40] Heyde, C.C. & Seneta, E. Estimation theory for growth and immigration rates in a multiplicative process. J. Appl. Prob. 9, 235-256 (1972).

[41] Heyde, C.C. & Seneta, E. Notes on "Estimation theory for growth and immigration rates in a multiplicative process". J. Appl. Prob. 11, 572-577 (1974).

[42] Hoel, D.G. & Crump, K.S. Estimating the generation-time distribution of an age-dependent branching process. Biometrics 30, 125-135 (1974).

[43] Jagers, P. A general stochastic model for population development. Skand. Akt. 1969, 84-103 (1969).

[44] Jagers, P. A limit theorem for sums of random numbers of i i d random variables. In: Jagers, P. & Råde, L. (ed.) Mathematics and Statistics. Essays in honour of Harald Bergström. Göteborg, pp. 33-39, 1973.

[45] Jagers, P. Maximum likelihood estimation of the reproduction distribution in branching processes and the extent of disintegration in cell proliferation. Tech. rep. 1973-17, Dep. of Math., Chalmers Univ. of Techn. and Univ. Göteborg, 1973.

[46] Jagers, P. Branching processes with biological applications. Wiley, New York, 1975.

[47] Jagers, P. & Norrby, K. Estimation of the mean and variance of cycle times in cinemicrographically recorded cell populations during balanced exponential growth. Cell Tissue Kinet. 7, 201-211 (1974).

[48] Kashyap, R.L. & Rao, A.R. Dynamic stochastic models from empirical data. Academic Press, New York, 1976.

[49] Keiding, N. Estimation in the birth process. Biometrika 61, 71-80 and 647 (1974).

[50] Keiding, N. Estimation theory for branching processes. Bull. Int. Statist. Inst. 46 (4), 12-19 (1975).

[51] Keiding, N. Maximum likelihood estimation in the birth-and-death process. Ann. Statist. 3, 363-372, (1975). Correction note Vol. 6, No. 2 (1978).

[52] Keiding, N. Contribution to the discussion of Efron [25].

[53] Keiding, N. Population growth and branching processes in random environments. Proc. 9 Int. Biom. Conf. Inv. Papers II, 149-165 (1976).

[54] Keiding, N. & Lauritzen, S. Marginal maximum likelihood estimates and estimation of the offspring mean in a branching process. Scand. J. Statist. 5 (to appear 1978).

[55] Kendall, D.G. Les processus stochastiques de croissance en biologie. Ann. Inst. H. Poincaré 13, 43-108 (1952).

[56] Kendall, D.G. Branching processes since 1873. J. London Math. Soc. 41, 385-406 (1966).

[57] Kesten, H. & Stigum, B.P. Additional limit theorems for indecomposable multidimensional Galton-Watson processes. Ann. Math. Statist. 37, 1463-1481 (1966).

[58] Lotka, A.J. Théorie analytique des associations biologiques. Actualités Sci. Indust. 2, no. 780, 123-136 (1939).

[59] Miller, R.S. Botkin, D.B. & Mendelssohn, R. The whooping crane (Grus americana) population of North America. Biol. Conserv. 6, 106-111 (1974).

[60] Moran, P.A.P. Estimation methods for evolutive processes. J. Roy. Statist. Soc. B 13, 141-146 (1951).

[61] Moran, P.A.P. The estimation of parameters of a birth and death process. J. Roy. Statist. Soc. B 15, 241-245 (1953).

[62] Nagaev, A.V. On estimating the expected number of direct descendants of a particle in a branching process. Theor. Prob. Appl. 12, 314-320 (1967).

[63] Pakes, A.G. Nonparametric estimation in the Galton-Watson process. Math. Biosc. 26, 1-18 (1975).

[64] Papangelou, F. Integrability of expected increments of point processes and a related change of scale. Trans. Amer. Math. Soc. 165, 483-506 (1972).

[65] Quine, M.P. Asymptotic results for estimators in a subcritical branching process with immigration. Ann. Prob. 4, 319-325 (1976) Corr. Note Vol. 5, 318 (1977).

[66] Quine, M.P. & Durham P. Estimation for multitype branching processes. Manuscript, University of Sydney, 1977.

[67] Rao, C.R. Linear Statistical Inference and its applications. Wiley, New York, 1965.

[68] Rudemo, M. On a random transformation of a point process to a Poisson process. In: Jagers, P. & Råde, L. (ed.) Mathematics and Statistics. Essays in honour of Harald Bergström. Göteborg, pp. 79-85, 1973.

[69] Stigler, S.M. Estimating the age of a Galton-Watson branching process. Biometrika 57, 505-512 (1970).

[70] Thompson, E.A. Estimation of age and rate of increase of rare variants. Am. J. Hum. Genet. 28, 442-452 (1976).

[71] Thompson, E.A. Estimation of the characteristics of rare variants. In: Measuring selection in natural populations. (F.B. Christiansen and T.M. Fenchel, ed.) Springer Lecture Notes in Biomathematics 19, 531-543 (1977).

[72] Yanev, N.M. On the statistics of branching processes. Theor. Prob. Appl. 20, 612-622 (1975).

MARTIN BOUNDARIES OF GALTON-WATSON PROCESSES

S. Dubuc

Université de Montréal
Montreal, Quebec, Canada

INTRODUCTION

In a remarkable paper, Martin [22] showed how to recover all positive harmonic functions on a domain Ω from minimal ones. The minimal harmonic functions themselves can be obtained as limits of quotients of Green functions. These quotients and their limits give a compactification of Ω, which is called Martin's compactification. This construction was used again by Naïm [23], and Brelot [4] in order to get deeper results in potential theory. In a similar way, boundaries to the state space of a discrete Markov chain were introduced by Feller [12], Doob [7], Hunt [15] and Watanabe [31] and are very useful for the analysis of regular functions and for the study of the asymptotic behavior of the chain. Martin's boundary due to Doob was retained by most of probabilists and was described for various particular Markov chains by the following people: Doob-Snell-Williamson [8], Lamperti-Snell [21], Ney-Spitzer [25], Hennequin [14] for random walks, Blackwell-Kendall [2] for Polya's urn scheme.

What we want to describe here is Martin's boundaries of a Galton-Watson process, entrance and exit boundary. Before we do so, we will summarize Choquet's theory of integral representation in compact convex sets and we will study the cone of non-negative invariant vectors of a given matrix of infinite order.

1. EXTREME POINTS THEORY

Let C be a convex cone of a locally convex Hausdorff topological vector space E, the null vector of E being the vertex of the cone C, one says that C has a compact basis if there is a continuous linear functional $L:E \to R$ such that $\forall x \in C-\{0\}$, $L(x) > 0$ and $K = \{x \in C:L(x) = 1\}$ is a compact set of E. If μ is a finite measure on a compact basis K of a convex cone C, there is one and only one point x of C such that for any L of E^*, $L(x) = \int_K L(y)\, d\mu(y)$ (see Phelps [26] for this result and others from this section). This point x is then called the resultant of the measure μ and the measure μ is also called an integral representation of x. Let us denote by $\mathcal{E}$ the set of extreme points of K, when K is a convex set. When K is metrizable, $\mathcal{E}$ is a G_δ, so it is a measurable subset of K. The following partial ordering on E is induced by $C:x \le y$ if and only if $y-x \in C$. C is said to be a lattice if any two elements of C has a smallest upper bound according to the previous ordering.

Theorem 1

(Choquet [5]) Let C be a convex cone with a metrizable compact basis K of a locally convex topological vector space E

and let $\mathcal{E}$ be the set of extreme points of K. For any point x of C, there is at least one integral representation μ of x which is supported by $\mathcal{E}$. If C is a lattice such a measure supported by $\mathcal{E}$ is unique.

A rather short proof of the existence of an integral representation supported by $\mathcal{E}$ was found by Bonsall [3]. Let us give a short proof of the uniqueness of the representation of a given vector x when C is a lattice. Let μ and ν be two representations of x supported by $\mathcal{E}$, let us use Hahn's decomposition of the signed measure $\nu-\mu$: there is a measurable subset A of such that $\nu-\mu$ is a positive measure on A and $\nu-\mu$ is a positive measure on $\mathcal{E}$-A. We get four measures: $\mu_1\mu_2,\nu_1$ and ν_2: $\mu_1(B) = \mu(A\cap B)$, $\mu_2(B) = \mu(B-A)$, $\nu_1(B) = \nu(A\cap B)$ and $\nu_2(B) = \nu(B-A)$. Let us denote by y_1,y_2,z_1 and z_2 the corresponding resultants of these four measures ; $x = y_1 + y_2 = z_1 + z_2$. According to the decomposition theorem in a vector lattice, (see [26], pp.61-62 for example), we find four vectors of C, $u_{11}, u_{12}, u_{21}, u_{22}$ such that $u_{11} + u_{12} = y_1$, $u_{21} + u_{22} = y_2$, $u_{11} + u_{21} = z_1$ and $u_{12} + u_{22} = z_2$. Since $\mu_1 \le \nu_1$, one gets $y_1 \ge z_1$, that is $u_{12} \ge u_{21}$. Since $\mu_2 \le \nu_2$, one gets $y_2 \le z_2$, that is $u_{12} \ge u_{21}$. Since $\mu_2 \le \nu_2$, one gets $y_2 \le z_2$, that is $u_{12} \le u_{21}$. So $u_{12} = u_{21}$ and $y_1 = z_1$. $\mu_1(\mathcal{E}) = \nu_1(\mathcal{E})$ and $\mu_1 = \nu_1$. Similarly $\mu_2 = \nu_2$ and $\nu = \mu$.

Lemma 2

Let K be a compact convex subset of a locally convex topological vector space E, if x is an extreme point of K and if F is a closed subset of K which does not contain x, then there is a continuous linear function L on E such that $\forall y \in F$ $L(y) \ge L(x) + 1$.

PROOF

For any y in F, one can find a closed convex neighborhood U_y of y which does not contain x. By compactness of F, there is a finite part S of F such that $F \subseteq \bigcup_{y \in S} U_y$. The convex hull W of $\bigcup_{y \in S} U_y \cap K$ is compact and since x is extreme, x does not belong to W. It is then possible to separate x and W by a continuous linear function; so the same thing is true for x and F.

Theorem 3

If $\{K_n\}$ is an increasing sequence of compact convex sets whose union is dense in a metrizable compact convex set K and if x is an extreme point of K, there is a sequence of points $\{x_n\}$ such that x_n is an extreme point of K_n and $x = \lim_{n \to \infty} x_n$.

PROOF

Let $\{O_n\}$ be a basis of neighborhoods of an extreme point x of K, by using again and again lemma 2, it is possible to find a sequence of continuous linear functions L_n such that the sequence $U_n = \{y\colon L_n(y) < L_n(x) + 1,\ y \in K\}$ has the following property

$$U_1 \subset O_1,\quad U_2 \subset O_2 \cap U_1,\quad U_3 \subset O_3 \cap U_2, \ldots$$

Since $\bigcup_{n=1}^{\infty} K_n$ is dense in K, there is an increasing sequence of integers $n_1, n_2, \ldots$ such that $K_{n_i} \cap U_i \neq \phi$. Let $\alpha_{in} = \inf\{L_n(y)\colon y \in K_n\}$ where $n \geq n_i$, there is an extreme point of K_n, x_{in} such that $L_n(x_{in}) = \alpha_{in}$. So $x_{in} \in U_i$. If $n_i \leq n < n_{i+1}$, we set $x_n = x_{in}$. $\{x_n\}$ is the required sequence.

Theorem 4

If F is a closed set whose convex hull is dense in a compact convex set K and if x is a point of F which is not extreme in K, then there is a probability measure μ on $F-\{x\}$ whose resultant is x.

PROOF

With the vague topology, $M^1(F)$, the set of probability measures on F, is convex and compact. For μ in $M^1(F)$, we note by

$r(\mu)$ the resultant of μ. The mapping $\mu \to r(\mu)$ is affine and continuous. The range of this application is then convex, compact. As it can be seen from lemma 2, any extreme point of K is in F. So any point of K is the resultant of a measure μ supported by F. If $x \in F$ and x is not extreme in K, it is not possible to find two distinct points y and z of K such that $x = (y+z)/2$. If σ and τ are two representing measures supported by F for y and z respectively, let $a = (\sigma\{x\} + \tau\{x\})$, $a < 2$. The measure $\mu = (\sigma + \tau - a\sigma_x)/(2-a)$, where σ_x is the unit mass at x, is a probability measure on $F-\{x\}$ which represents x.

2. INTEGRAL REPRESENTATION OF SUPERREGULAR FUNCTIONS

Let S be a countable set and $P = (p(x,y))$ a non negative matrix, $x \in S$, $y \in S$, $p(x,y) \geq 0$. The matrix P is not necessarily stochastic nor substochastic. The only other assumption is the finiteness of entries of powers of P: $P^n = (p_n(x,y))$, $n = 0,1,2,\ldots$ where $p_0(x,y) = \delta(x,y)$, Kronecker's symbol of x and y and $p_{n+1}(x,y) = \Sigma_{z\in S}\, p(x,z)\, p_n(z,y)$. We assume that $(\forall x \in S)\ (\forall y \in S)\ p_n(x,y) < \infty$. We introduce the cone of superregular functions: $\mathcal{S} = \{h: S \to (0,\infty) \mid \Sigma_{y\in S}\, p(x,y)h(y) \leq h(x)\}$. What we will prove in that section is (1) that we can find a part $\mathcal{E} = \{h(x,e)\}$ of $\mathcal{S}$ and a measurable structure on $\mathcal{E}$ such that for each choice of x the map $e \to h(x,e)$ is measurable and (2) for any superregular function h of $\mathcal{S}$, there is a unique measure μ on $\mathcal{E}$ such that $h(x) = \int h(x,e)\, d\mu(e)$ for all x of S. Partly, we will use Neveu's approach [24] which relies on Choquet's theorem.

The Green's function corresponding to the matrix P is by definition: $G(x,y) = \Sigma_{n=0}^{\infty} P_n(x,y)$; $0 \leq G(x,y) \leq \infty$. A potential is by definition a finite function $p(x)$ of the form $\Sigma_{y\in S}\, G(x,y)\cdot \psi(y)$ where $\psi \geq 0$. The collection of all potentials will be noted by $\mathcal{P}$. A regular function h is a function on S such that $Ph = h$

i.e. $\Sigma_{y \in S}\, p(x,y)\, h(y) = h(x)$. $\mathcal{H}$ will be the collection of non negative regular functions. We give what is known as Riesz's decomposition.

Theorem 5

If s is a superregular non negative function, then $s(x) = p(x) + h(x)$ where $p \in \mathcal{P}$, $h \in \mathcal{H}$. This decomposition is unique.

PROOF

See Kemeny, Snell and Knapp [18]. Here $\psi = s - Ps$, $p = G\psi$, $h = s - p$. h is also the monotone limit of $P^n s$.

Remark

It is sometimes useful to notice that $G\psi' = G\psi$ will happen only if $\psi = \psi'$. Indeed $PG\psi = PG\psi'$. But $G = I + PG$ where I is the identity matrix. So $I\psi = I\psi'$, $\psi = \psi'$.

Theorem 6

$\mathcal{S}$ is a lattice for the ordering it induces on itself.

PROOF

See Neveu [24].

If $p_1 = G\psi_1$, $p_2 = G\psi_2$, then $p_1 \wedge p_2 = G\psi$ where $\psi(x) = \min(\psi_1(x), \psi_2(x))$. If h_1 and h_2 are in H, $h_1 \wedge h_2$ is the regular function in Riesz's decomposition of the function $\min\,(h_1(x), h_2(x))$.

To go further in the analysis of $\mathcal{S}$, we introduce the following quantities $\{f_k(x,y)\}_{k=1}^{\infty}$, $F(x,y)$, $H_A(x,y)$. If $x \in S$, $A \subseteq S$, $y \in A$, we define by induction

$$f_1(x,y;A) = p(x,y)$$

$$f_{k+1}(x,y;A) = \sum_{z \notin A} p(x,z)\, f_k(z,y;A), \qquad k \geq 1$$

$$F_A(x,y) = \sum_{k=1}^{\infty} f_k(x,y;A)$$

$$H_A(x,y) = \begin{cases} \delta(x,y) & \text{if } x \in A \\ F(x,y;A) & \text{if } x \notin A \end{cases}$$

If A has only one point $\{y\}$, we write $f_k(x,y)$, $F(x,y)$ and $H(x,y)$ instead of $f_k(x,y;A)$, $F_A(x,y)$ and $H_A(x,y)$. When P is a stochastic matrix, $f_k(x,y)$ is the probability of a first visit at y at time k starting at x. F and H_A has similar probabilistic interpretation. (See Spitzer [30]).

Theorem 7

If h is a superregular function, then $h(x) \geq \Sigma_{y \in A} F_A(x,y) \cdot h(y)$.

If moreover $h(x_0) = \Sigma_{y \in A} F_A(x_0,y)\, h(y)$ at some point x_0 of S, then $h(x_0) = \Sigma_{y \in S}\, p(x_0,y)\, h(y)$.

PROOF

Let us define a sequence of rectangular matrix $p_n'(x,y';A)$ where $x \in S$, $y' \notin A$:

$$p_1'(x,y';A) = p(x,y')$$

$$p_{n+1}'(x,y';A) = \sum_{z \notin A} p_n'(x,z)\, p(z,y').$$

Let h be a superregular function. By induction, the following sequence is decreasing

$$\sum_{k=1}^{n} \sum_{y \in A} f(x,y;A)\, h(y) + \sum_{y' \notin A} p_n'(x,y';A)\, h(y')$$

So $\Sigma_{k=1}^{n} \Sigma_{y \in A} f_k(x,y;A)\, h(y)$ is bounded by $\Sigma_{y \in S}\, p(x,y)\, h(y')$ and $\Sigma_{y \in A} F_A(x,y)\, h(y) \leq h(x)$. If $\Sigma_{y \in A} F_A(x_0,y)\, h(y) = h(x_0)$, then $\Sigma_{y \in S} p(x_0,y)\, h(y) = h(x_0)$.

It is possible to find a (finite or infinite) sequence of points of S, $\{x_n\}$ $n \geq 1$ and a sequence of subsets $\{S_n\}$ $n \geq 1$ such that

(a) $x_n \notin S_{n-1}$

(b) $S_n = S_{n-1} \cup \{y \mid \exists k \geq 0 \; p_k(x_n,y) > 0\}$

(c) $\bigcup_n S_n = S$

If $\sigma_n = S_n - S_{n-1}$ $(S_0 = \phi)$, we will say that σ_n is an admissible covering of S with x_n as vertices.

Theorem 8

If $\{\sigma_n\}$ is an admissible covering of S with x_n as vertices, if $h(x)$ is a superregular function of $\mathcal{S}$ then there is a sequence $\{h_n\}$ of superregular functions of S such that

(a) $h(x) = \Sigma_n h_n(x)$

(b) $h_n(x) = 0$ if $x \in \sigma_1 \cup \sigma_2 \cup \ldots \cup \sigma_{n-1}$

(c) $h_n(x) = \Sigma_{y\in\sigma_n} H_{\sigma_n}(x,y)\, h_n(y)$.

PROOF

$h_1(x) = \Sigma_{y\in\sigma} H_{\sigma_1}(x,y)\, h(y)$. Since for $x \notin \sigma_1, \Sigma_{y\in\sigma_1} p(x,y)\cdot H_{\sigma_1}(y,z) = H_{\sigma_1}(x,z)$ then $\Sigma_{y\in S}\, p(x,y)\, h_1(y) = h_1(x)$ if $x \notin \sigma_1$ according to theorem 7, $h(x) \geq h_1(x)$. So $h(x) - h_1(x)$ belongs to $\mathcal{S}$ and this function is null on σ_1. By recurrence, one defines

$$h_n(x) = \sum_{y\in\sigma_n} H_{\sigma_n}(x,y)\, r_{n-1}(y),$$

where $r_{n-1}(x) = h(x) - (h_1(x)+\ldots+h_{n-1}(x))$. $\{h_n\}$ has desirable properties. Finally if x is a given point of σ_1, then

$$h(x) = \sum_{n=1}^{i} h_n(x) = \sum_n h_n(x).$$

Definition

In a convex cone C, we say that a vector h of C is *extreme* if it is impossible to express h as the sum of two linearly independent vectors of C.

Theorem 9

It is possible to find a collection of extreme superregular functions $\mathcal{E} = \{h(x,e)\}$ and a measurable structure on $\mathcal{E}$ such that

(a) any extreme superregular function is a scalar multiple of a function of $\mathcal{E}$,

(b) for any x of S, the mapping $e \to h(x,e)$ is measurable,

(c) for any function h of S, there is one and only one measure μ on $\mathcal{E}$ such that

$$h(x) = \int h(x,e)\, d\mu(e).$$

PROOF

If $\{\sigma_n\}$ is an admissible covering of S with $\{x_n\}$ as vertices, we introduce the subcones

$$\mathcal{S}_n = \{h \in \mathcal{S}: h(x) = \sum_{y \in \sigma_n} H_{\sigma_n}(x,y)\, h(y)\}$$

and

$$C_n = \{h: \sigma_n \to [0,\infty] \mid \sum_{y \in \sigma_n} p(x,y)\, h(y) \le h(x)\}.$$

Let us use on C_n the topology of simple convergence. Let us check that C_n is a linear convex cone with a compact base. Let $L_n h = h(x_n)$, this is a continuous linear function. If $K_n = \{h \in C_n: h(x_n) = 1\}$ and if $y \in \sigma_n$, there is an integer m such that $p_m(x_n,y) > 0$; so if $h \in K_n$, $h(y) \le \frac{1}{P_m(x_n,y)} = c(y)$. K_n is contained in the compact set

$$\{h: \sigma_n \to (0,\infty) \mid h(y) \le c(y)\},$$

K_n is closed for the topology of simple convergence, K_n is compact. According to theorem 6, C_n is a lattice. Let $\mathcal{E}_n^a$ be the set of extreme points of K_n. If $x \in S$, the mapping $h \to \Sigma_{y \in \sigma_n} H_{\sigma_n}(x,y) \cdot h(y)$ is semi-continuous.

So $\mathcal{E}_n = \{h \in \mathcal{E}_n^a: (\forall x \in S)\ \Sigma H_{\sigma_n}(x,y)\, h(y) < \infty\}$ is a measurable set. Any function h of $\mathcal{S}_n$ has a unique integral representation by a measure supported by $\mathcal{E}_n$: $h(x) = \int_{\mathcal{E}_n} h(x,e)\, d\mu_n(e)$.

One can use $\mathcal{E} = \cup_n \mathcal{E}_n$ as a set which gives unique integral representation for functions h of $\mathcal{S}$.

3. MARTIN SPACE OF A NON NEGATIVE MATRIX

We will explain in that section that any extreme vector of S can be computed as limit of the function $F(x,y)/F(a,y)$. We now study properties of the functions $F(x,y)$ as defined in the previous section. The first thing to observe is the equation

$$\sum_{y \in S} P(x,y)\ F(y,a) + p(x,a)(1-F(a,a)) = F(x,a).$$

Let $S_0 = \{a\colon F(a,a) \le 1 \text{ and } \forall x \in S,\ F(x,a) < \infty\}$, the Martin space $\mathcal{M}(P)$ of the matrix P will be by definition the closure for the topology of simple convergence on S of the set of functions $\{\lambda F(x,a)\colon \lambda \in (0,\infty),\ a \in S_0\}$. When P is stochastic, one can compare with other definitions like that one of Derriennic [6], $\mathcal{S}_0$ will be finite linear combinations with non negative coefficients of $\{F(x,a)\colon a \in S_0\}$. We see that $\mathcal{S}_0$ and $\mathcal{M}(P)$ are parts of $\mathcal{S}$. We will show that $\mathcal{S}_0$ is dense in $\mathcal{S}$ and that any extreme vector of $\mathcal{S}$ belongs to $\mathcal{M}(P)$.

Theorem 10

If $F(a,a) \ge 1$, $F(x,a) > 0$, then $G(x,a) = \infty$; if $F(a,a) < 1$ and if $x \ne a$, then $G(a,a) = (1-F(a,a))^{-1}$ and $G(x,a) = F(x,a)/(1-F(a,a))$.

PROOF

As for a stochastic matrix $p(x,y)$, one uses the formulas

$$p_n(x,a) = \sum_{k=1}^{n} f_k(x,a)\ p_{n-k}(a,a).$$

Lemma 11

If $a \in S_0$, $F(x,a)$ as function of x is extreme in $\mathcal{S}$.

PROOF

If $F(a,a) < 1$, $F(x,a) = 1-F(a,a))\, G(x,a)$; $G(x,a)$ is extreme as it follows from Riesz's decomposition, the same is true for $F(x,a)$. If $F(a,a) = 1$, let h_1 and h_2 be two functions in $\mathcal{S}$ such that $F(x,a) = h_1(x) + h_2(x)$. From theorem 7, it follows that

$$h_1(x) \geq F(x,a)\, h_1(a) \quad \text{and} \quad h_2(x) \geq F(x,a)\, h_2(a)$$

$$F(x,a) = h_1(x) + h_2(x) \geq F(x,a)(h_1(a) + h_2(a)) = F(x,a)$$

$$h_1(x) = h_1(a)\, F(x,a) \quad \text{and} \quad h_2(x) = h_2(a)\, F(x,a).$$

Theorem 12

If $F(a,a) = 1$ and if $0 < F(a,b) < \infty$, then

$$F(b,b) = F(a,b)\, F(b,a) = 1.$$

PROOF

Let $A = \{a,b\}$, the following formulas are true:

$$f_k(a,a) = f_k(a,a;A) + \sum_{i=1}^{k-1} f_i(a,b;A)\, f_{k-i}(b,a)$$

$$f_k(a,b) = f_k(a,b;A) + \sum_{i=1}^{k-1} f_i(a,a;A)\, f_{k-i}(a,b)$$

$$F(a,a) = F(a,a;A) + F(a,b;A)\, F(b,a) \tag{1}$$

$$F(a,b) = F(a,b;A) + F(a,a;A)\, F(a,b). \tag{2}$$

If $F(a,a) = 1$, from (1), it follows that $F(a,a;A) = 1-F(a,b;A) \cdot F(b,a)$; substituting this value of $F(a,a;A)$ in (2), we get $F(a,b)\, F(b,a) = 1$.

$$F(b,a) = F(b,a;A) \quad F(b,a;A)\, F(b,a) \tag{3}$$

$$F(b,b) = F(b,b;A) \quad F(b,a;A)\, F(a,b). \tag{4}$$

If we replace in (4) the value of $F(b,a;A)$, $F(b,a;A) = F(b,a) - F(b,b;A)F(b,a)$, $F(b,b) = 1$.

Theorem 13

If $F(a,a) = 1$, if $F(a,b) > 0$ and if $h \in \mathcal{S}$, then $h(b) = F(b,a)\, h(a)$.

PROOF

From Theorem 7, it follows that

$$h(b) \geq F(b,a)\, h(a) \quad \text{and} \quad h(a) \geq F(a,b)\, h(b).$$

So $h(b) \geq F(b,a)\, h(a) \geq F(b,a)\, F(a,b)\, h(b)$. From theorem 12, $F(a,b)\, F(b,a) = 1$, so $h(b) = F(b,a)\, h(a)$.

Theorem 14

$\mathcal{S}_0$ is dense in $\mathcal{S}$ if one uses simple convergence over S.

PROOF

Let h be a function of $\mathcal{S}$. The first remark is that $h(x) = 0$ if $x \notin S_0$, this follows easily from theorem 7. Let $R = \{a \in S_0 : F(a,a) = 1\}$, we set $h_0(x) = \Sigma_{y \in R}\, F_R(x,y)\, h(y)$ and $h_1(x) = h(x) - h_0(x)$. h_0 is regular $(Ph_0 = h_0)$ and $h_1 \in \mathcal{S}$. Let $a_1, a_2, \ldots, a_n$ be n points of S_0 such that $F(a_i, a_i) < 1$, we set $G_1(x) = \Sigma_{i=1}^{n}\, h_1(a_i)\, G(x_1, a_1)$ and $h_2(x) = \min(h_1(x), G_1(x))$. h_2 is a potential and $h_2(a_i) = h_1(a_i)$. If B is a finite set of S such that $B \cap S_0 - R = \{a_1, a_2, \ldots, a_n\}$, it is possible to approach $h_2(x)$ on B by non negative linear combinations of $\{F(x,a) : a \in S_0 - R\}$ and $h_0(x)$ on B by non negative linear combinations of $\{F(x,a) : a \in R\}$.

Theorem 15

Any extreme function of $\mathcal{S}$ belongs to the Martin space of P; more precisely, if $h(x)$ is an extreme function of $\mathcal{S}$, if $h(a) \neq 0$, there is a sequence of points a_n such that for any x of S

$$\frac{h(x)}{h(a)} = \lim_{n\to\infty} \frac{F(x\ ,a_n)}{F(a\ ,a_n)} .$$

PROOF

If h_0 is an extreme function of $\mathcal{S}$ and if h_0 is not identical to $\lambda F(x,a)$ for some $\lambda \geq 0$ and $a \in S_0$, then h_0 is regular. If $\{\sigma_n\}$ is an admissible covering of S with $\{x_n\}$ as vertices, there will be a point $a = x_n$ such that $h_0(a) \neq 0$. Let $B = \{y: y=a \text{ or } F(y,a) > 0\}$, if $K = \{h: B \to (0,\infty), h(x) \geq \Sigma_{y\in B}\, p(x,y)\, h(y), h(a) = 1\}$, K is compact, convex. According to theorem 14, the convex hull of $\{F(x,b)/F(a,b): b \in B\}$ is dense in K. It follows from theorem 3 that there is a sequence of points a_{n0} of B such that $h_0(x)/h_0(a) = \lim_{n\to\infty} [F(x,a_{n0})/F(a,a_{n0})]$. One concludes the proof by showing that $\lim_{n\to\infty} [F(x,a_n)/F(a,a_n)] = h_0(x)/h_0(a)$ even if $x \notin B$ for some sequences a_n of B by using a diagonal argument.

4. MARTIN BOUNDARIES OF A GALTON-WATSON PROCESS

Let $f(s) = \Sigma_{k=0}^{\infty} p_k s^k$ be the generating function of the offspring distribution of a simple Galton-Watson process. Let $\{Z_n\}_{n=0}$ be the corresponding Markov chain and $p(x,y) = \Pr(Z_1 = y \mid Z_0 = x)$. We will describe Martin boundaries arising from this matrix P. To avoid technical details, we will assume that the process is indecomposable (1 is the only common divisor of $\{k\text{-}m: p_k \neq 0, p_m \neq 0\}$ and is of finite variance, $\Sigma k^2 p_k < \infty$.

(a) Extreme harmonic function. If one looks for extreme function such that $h(x) = \Sigma_{y=0}^{\infty} p(x,y)\, h(y)$, one such candidate is $h(x) = F(x,0) = q^x$, q being the probability of extinction of the process when $Z_0 = 1$. Indeed 0 is an absorbing state and is the only recurrent element. To get the other extreme harmonic function, one can study limits of $F(x,y)/F(1,y) = G(x,y)/G(1,y)$ as $y \to \infty$. In the critical case, $m = \Sigma_{k=1}^{\infty} kp_k = 1$, Kesten-Ney and Spitzer [9] showed that $\lim_{y\to\infty} [G(x,y)/G(1,y)] = x$ under the condition $\Sigma k^2 \log k \; p_k < \infty$. In the critical case, the only other extreme harmonic function is $h(x) = x$ (see also Dubuc, Proc.Amer. Math. Soc. 21 (1969), 324). In the supercritical case, the study of harmonic functions is closely related to limit law $W = \lim_{n\to\infty} Z_n m^{-n}$ where $Z_0 = 1$, $m = \mathcal{E}(Z_1)$. Let W_x be the sum of x independent copies of W and let $w_x(t)$ the density of W_x for $t > 0$. If y_n is a sequence of integers such that $y_n m^{-n}$ converge to $c \in (0,\infty)$, then $\lim_{n\to\infty} y_n p_n(x,y_n) = xw_x(c)$ (see Dubuc-Seneta [11]). One has also $\lim_{n\to\infty} y_n G(x,y_n) = \Sigma_{n=-\infty}^{\infty} cm^n w_x(cm^n) = h(x;c)$ (cf. Dubuc [9] or Athreya-Ney [1]). Using oscillatory behavior of $h(x,c)$ when x tends to ∞ and using theorem 4, one can show that each $h(x,c)$ is extreme (see [10]). The subcritical case can be reduced to the supercritical one in some situations. If $f(s)$ can be extended analytically on a disk $\{z: |z| \le R\}$ such that $f(R) = R$, $f'(R) < \infty$, $R > 1$ and if $f_*(s) = f(sR)/R$, let $p_*(x,y)$ the corresponding matrix for the Galton-Watson Z_n^*, then $p_*(x,y) = p(x,y)R^{y-x}$. There will be a one-to-one correspondence between superregular functions for p and those for p^*, the mapping being $h_*(x) = h(x)\, R^{-x}$.

(b) Space-time harmonic function. If $p(x,y)$ are the transition probabilities of a given Galton-Watson process, on $\mathbb{N}\times\mathbb{Z}$, we define the matrix

$$\bar{p}(x,t,y,t+1) = p(x,y) \qquad x \in \mathbb{N}, \quad y \in \mathbb{N}, \quad t \in \mathbb{Z}$$

$$\bar{p}(x,t,y,u) = 0 \qquad \text{if} \qquad u \neq t+1.$$

If $\bar{h}(x,t) = cm^t\, w_x(cm^t)$, $\bar{h}$ is a space-time harmonic as was noted by Athreya-Ney [1]. In a forthcoming paper (Ann.Inst.Henri Poincare), Lootgieter will study these functions and will solve various problems with them.

(c) Space-time harmonic measure. If one looks for a sequence of measures on $\mathbb{N}$, $\{\mu_n\}_{n=-\infty}^{\infty}$ such that $\mu_{n+1}(x) = \Sigma_{x\in N}\, \mu_n(x)p(x,y)$, what we will call an extrance law, this comes to find regular function for the matrix

$$\bar{p}^*(x,n,y,n-1) = p(y,x)$$

$$\bar{p}^*(x,n,y,m) = 0 \qquad m \neq n-1.$$

For a subcritical Galton-Watson process, Joffe-Spitzer [16] described the following entrance laws: if $g(s)$ is the generating function of the Yaglom's limit law, the Poisson compounds, μ_n, whose generating functions are $\exp(-Am^n(1-g(s)))$, give entrance law for the Galton-Watson process. Since these measures are limits of Green's functions, it seems that these laws are extreme.

(d) Invariant measures. There is just one truly invariant measure for a given Galton-Watson process (see Harris [13] for example). If one forgets the state 0, one looks for non negative solution to the set of equations $\mu(x) = \Sigma_{y=1}^{\infty}\, \mu(y)\, p(y,x)$, $x \geq 1$. In the critical case, Kesten, Ney and Spitzer [19] proved that invariant measures are unique up to a multiplicative constant. Karlin and McGregor [17] studied the behavior of $\mu(x)$ when x goes to ∞. When the process is not critical, Kingman [20] showed that invariant measures are not necessarily unique. In the subcritical situation, Seneta [29] showed that amongst all invariant measures there is just one which satisfies a regular variation condition. Hoppe extended this result to multitype branching process. (Journal of Applied Probability, 1976) In a forthcoming paper, Hoppe will describe the Martin boundary of these invariant measures (Representations of invariant measures on multitype Galton-Watson processes, technical report, Université de Montréal, 1975, 7 pages).

In conclusion, we can say that many things are known about Martin boundaries of a Galton-Watson process. However, there are many open questions either for this class of processes or for a more general branching process as multitype branching processes or branching process with immigration as did study Spitzer, Seneta or Hoppe. Another question is the usefulness of these boundaries.

Savits [27,28] was able to analyze the growth behavior of the process from space-time harmonic functions. This is one application.

REFERENCES

1. K.B. Athreya and P.Ney, Branching Processes, Springer-Verlag, New York, 1972.
2. D.Blackwell and D.Kendall, J.Appl.Probability,1:284 (1964).
3. F.F.Bonsall, J.London Math.Soc.,38:332 (1963).
4. M.Brelot, J.Math. Pures Appl., 35: 297 (1956).
5. G.Choquet, Seminaire Bourbaki, décembre 139 (1956).
6. Y.Derriennic, Ann.Inst.Henri Poincare B 9: 233 (1973).
7. J.L.Doob, J.Math.and Mech.,8: 433 (1959).
8. J.L.Doob, J.L.Snell and R.E.Williamson, in Contributions to to Probability and Statistics, 1960.
9. S.Dubuc, Studia Math. 34: 69 (1970).
10. S.Dubuc, Ann.Inst.Fourier, 21: 171 (1971).
11. S.Dubuc and E.Seneta, Ann.Probability,4:490 (1976).
12. W.Feller, Trans.Amer.Math.Soc., 83: 19 (1956).
13. T.E.Harris, The Theory of Branching Processes, Springer-Verlag, Berlin, 1963.
14. P.L.Hennequin, Ann.Inst.Henri Poincare 2, 18: 109 (1963).
15. G.A.Hunt, Illinois J.Math. 4: 313 (1960).
16. A.Joffe and F.Spitzer, J.Math.Anal.Appl., 19: 409 (1967).
17. S.Karlin and J.McGregor, Ann.Math.Statist., 38:977 (1967).
18. J.Kemeny, L.Snell and A.Knapp, Denumerable Markov Chains, van Nostrand, Princeton, 1966.
19. H.Kesten, P.Ney and F.Spitzer, Teor.Verojatnost. i Primenen , 11: 579 (1966).
20. J.F.C.Kingman, Proc.Amer.Math.Soc., 16: 245 (1965).
21. J.Lamperti and J.L.Snell, J.Math.Soc.Japan, 15:113 (1963).
22. R.S.Martin, Trans.Amer.Math.Soc. 49: 137 (1941).
23. L.Naim, Ann.Inst.Fourier, 7: 183 (1957).

24. J.Neveu, Ann.Fac.des Sciences de Clermont, 24: 37 (1964).

25. P.Ney and F.Spitzer, Trans.Amer.Math.Soc., 121:116 (1966).

26. R.R.Phelps, Lectures on Choquet's Theorem, van Nostrand, Princeton, 1966.

27. T.H.Savits, Ann.Probability, 3: 61 (1975).

28. T.H.Savits, Adv.Appl.Prob.,7:283 (1975).

29. E.Seneta, J.Appl.Prob., 8: 43 (1971).

30. F.Spitzer, Principles of Random Walk, van Nostrand, Princeton 1964.

31. T.Wanatabe, Mem.Coll.Sc.,U.of Kyoto, Ser.A, 33: 39 (1960).

INVARIANCE PRINCIPLES FOR THE SPATIAL DISTRIBUTIONS OF BRANCHING POPULATIONS

Luis G. Gorostiza*

Centro de Investigación del Instituto Politécnico Nacional
Mexico D.F., Mexico

Alberto Ruíz Moncayo**

Universidad Autónoma Metropolitana Iztapalapa
Mexico D.F., Mexico

1. INTRODUCTION

The general model with which we are concerned is a population evolving as a supercritical branching process, with or without age dependence, where the offspring move at random in space according to certain laws, and the question is what can be said about the distribution of their positions at a given time, as this time goes to infinity. Typical models that have been studied by several au-

---

* Partially supported by CONACYT Grant PNCB 000060

** Partially supported by CONACYT Grant PNCB 10010802

thors are branching random walks, and branching diffusions. In the former case the offspring jump when they are born, and retain their positions until they die or reproduce. In the latter case the offspring move throughout their lives according to diffusions. In both models the offspring motions are independent of each other and identically distributed, and independent of the family tree.

The empirical distribution of the positions of the offspring that are alive at a given time is defined by selecting the offspring with equal probability (the reciprocal population size at that time), and registering their positions. The empirical distribution is a random measure, because it depends on realizations of the branching process and of the offspring motions. Then, there are two questions: does the random empirical distribution of the offspring positions at a given time converge weakly (i.e., in the sense of convergence of distributions) to a non-random distribution, as the time tends to infinity, and how strongly, with respect to the basic probability space, does this convergence take place? There are various positive answers to these questions.

We will call such convergence results as regards the above questions central limit theorems, independently of the strength of convergence on the basic probability space, because they refer to the asymptotic behavior of a distribution (the empirical spatial distribution) attached to a single given point in time, as this point in time goes to infinity. But it is also of interest from a theoretical point of view, as well as for applications, to study not only the empirical distribution of the offspring positions at a given time, but more generally, the empirical distribution of the trajectories in space of the ancestry lines of the offspring alive at that time. And in the same way as the invariance principle extends the central limit theorem to a space of trajectories, it is natural to expect that the central limit theorems for the empirical spatial distributions might be extended to respective invariance principles, or functional central limit theorems. Questions that arise in applications often involve asymptotic distributions of functionals of the whole spatial trajectories of the descent lines, such as maximal displacement, first passage times, time spent in a set; thus the relevance of invariance principles in applied branching process theory.

Central limit theorems have been proven for some time, with increasing generality and diversity of the underlying models, and technical refinements. We will not survey these results here, but will mention some of them briefly, as motivation for the invariance principles that we propose to discuss. The object of this paper is to present a concise report on the present status and possible extensions of such invariance theorems.

The paper is organized in the following way. In Section 2 we review briefly some of the central limit theorems. Section 3 contains almost-sure invariance principles, and strong laws of large numbers that have been proven for certain cases. And in Section 4 we mention possible extensions of the previous results.

We will use freely basic results from the theory of branching processes, that are in the book of Athreya and Ney [7], and also weak convergence theory, from the book of Billingsley [8].

2. CENTRAL LIMIT THEOREMS

We consider supercritical Galton-Watson, and Bellman-Harris (or age-dependent) populations. For simplicity we make the unessential assumptions that the population starts out with one initial parent, and that each member generates at least one offspring; in the age-dependent case, we assume that the offspring lifetime distribution is non-lattice, without atom at 0, and the initial parent is of age 0 at time 0. On the population structure we superimpose random motions in space that the offspring perform throughout their lives. A basic probability space where everything is defined is assumed to exist.

In a branching random walk, the offspring motions are such that if a parent is located at x, then an offspring of this parent jumps at birth to $x+Y$, and remains at this position throughout its life, with the offspring jumps Y being all independent and identically distributed, and independent of the family tree. The process defined by the sequence of offspring motions along any given individual branch is a random walk. We suppose for simplicity that the motion is one-dimensional.

In the Galton-Watson case, let

ξ_n = the size of the n-th generation,

$\xi_n(A)$ = the number of elements in the n-th generation located in the Borel set A.

Then the proportion $\xi_n(\cdot)/\xi_n$ of members of the n-th generation located in Borel sets is a random measure. Assuming that the offspring jumps have mean μ and finite variance σ^2, Harris [14] (p. 75) conjectured that for real x,

$$\xi_n((-\infty, n\mu + n^{1/2}\sigma x])/\xi_n \to \Phi(x) \quad \text{in probability}$$

as $n\to\infty$, where Φ is the standard gaussian distribution function. Under the offspring production moment condition $E\xi_1^2 < \infty$, this was settled affirmatively by Ney [22]. Joffe and Moncayo [17] proved that the result holds with almost-sure (a.s.) convergence. Stam [25] also proved a.s. convergence, using a different method, and allowing offspring jumps with infinite variance and a stable distribution instead of Φ. Kaplan and Asmussen [19] reduced the moment condition $E\xi_1^2 < \infty$, and Kaplan [18] has shown that a.s. convergence holds under the minimum moment condition $E\xi_1 \log \xi_1 < \infty$.

For the Bellman-Harris model, let

Z_t = the size of the population at time t,

$Z_t(A)$ = the number of elements alive at time t located in the Borel set A.

The problem is also to prove convergence of the random measure

$Z_t(\cdot)/Z_t$ to a non-random distribution as $t \to \infty$. For offspring jumps with finite variance, mean-square convergence is shown in [7] (p. 238) under $E\xi_1^2 < \infty$, and Kaplan [18] proves a.s. convergence under $E\xi_1 \log \xi_1 < \infty$, in a theorem that includes the distribution of the offspring ages. Considering only the offspring positions, the result is that for each x,

$$Z_t((-\infty, t\mu\mu_\alpha^{-1} + t_\alpha^{1/2}(\sigma^2\mu_\alpha^{-1} + \mu^2\sigma_\alpha^2\mu_\alpha^{-3})^{1/2}x])/Z_t \to \Phi(x) \text{ a.s.}$$

as $t \to \infty$, where μ and σ^2 are the offspring jump mean and variance, and μ_α and σ_α^2 are the mean and variance of the distribution

$$\tilde{G}(t) = m \int_0^t e^{-\alpha y} dG(y), \quad t \geq 0,$$

where G is the offspring lifetime distribution, m is the mean of the offspring production law, and α is the Malthusian parameter, defined by

$$m \int_0^\infty e^{-\alpha y} dG(y) = 1.$$

In a <u>branching Brownian motion</u>, the population evolves according to the Bellman-Harris model, and the offspring perform Brownian motions throughout their lives, independently of each other and of the family tree, each offspring starting from the location of its birth. Again we assume that the motion is one-dimensional. The result is that under $E\xi_1^2 < \infty$,

$$Z_t((-\infty, t^{1/2}x])/Z_t \to \Phi(x) \text{ a.s.}$$

as $t \to \infty$. This was first shown by S. Watanabe in the case of exponential life distribution, and by Kaplan and Asmussen [19] in general. (The latter authors mention in [19] that the ξ_1 moment condition may be weakened for certain life distributions).

We have mentioned only the results that are more relevant for the following sections. Other works on branching diffusions and related models are [1], [2], [3], [4], [9], [10], [15], [26], [27]. Earlier work may be found in the references of the papers we have cited.

In the results that we have given explicitly, the quantity considered is the proportion of offspring located in a set, and in order to obtain a limit, a normalization is applied to the set. Some of the above mentioned papers also contain limit theorems where the set remains unchanged, and the proportion itself is normalized; in this case, the results we have stated have counterparts where the limit is the Lebesgue measure.

3. INVARIANCE PRINCIPLES AND LAWS OF LARGE NUMBERS

The results stated in the previous section are limits of the empirical spatial distribution (as defined in the Introduction) of the offspring in generation n, or at time t. We now write these results in the following more convenient way: In the age-dependent case,

$$Z_T^{-1}\Sigma_{\gamma\in\Gamma_T} 1[a_T^{-1}\pi_T(\gamma)\in A]\to\Phi(A) \quad \text{a.s.}$$

as $T\to\infty$, where Φ is interpreted here as the standard Gaussian measure, A is a Borel set of the real line, Γ_T denotes the set of branches of time-length T, the random variable $\pi_T(\gamma)$ is the centered position of the offspring on the branch γ at time T, a_T is the normalization, and $1[\ ]$ is the indicator of the set $[\ ]$. For the Galton-Watson case, T is replaced by n, Z_T by ξ_n, Γ_T by Γ_n (the branches up to generation n), etc.

In particular, for the Galton-Watson branching random walk, assuming for simplicity that the offspring jumps have mean 0 and variance 1, we have

$$\xi_n^{-1}\Sigma_{\gamma\in\Gamma_n} 1[n^{-1/2}\,\Sigma_{j=1}^{n} Y_j(\gamma)\in A]\to\Phi(A) \quad \text{a.s.}$$

as $n\to\infty$, where $Y_j(\gamma)$ is the jump of the j-th offspring on the branch γ. Now we scale the time, and introduce the process $X_n(\gamma)$ corresponding to $\gamma\in\Gamma_n$ as follows: the value of $X_n(\gamma)$ at time k/n is $\Sigma_{j=1}^{k} Y_j(\gamma)$, for $k = 0,1,\ldots,n$, with linear interpolations in between. $X_n(\gamma)$ is a random element of $C[0,1]$, the space of real continuous functions on [0,1] with the sup-norm topology. A special case of a theorem in [12] is that under $E\xi_1^2 < \infty$,

$$\xi_n^{-1}\Sigma_{\gamma \varepsilon \Gamma_n} 1[n^{-1/2} X_n(\gamma) \varepsilon A] \to B(A) \quad \text{a.s.}$$

as $n \to \infty$, where B is the standard Wiener measure, and A is a Borel set of C[0,1] with boundary of B-measure 0. This a.s. invariance principle extends the above a.s. central limit theorem in the same way as Donsker's theorem extends the classical central limit theorem.

For the branching Brownian motion with exponential lifetime distribution, let $X_T(\gamma)$ denote the process corresponding to the branch $\gamma \varepsilon \Gamma_T$, whose value at time $t \varepsilon [0,1]$ is the position of the offspring that lives on γ at time Tt. Hence $X_T(\gamma)$ is a Brownian random element of C[0,1], because by piecing the offspring Brownian motions along a branch one obtains Brownian motion again. In [13] we have shown that under $E\xi_1^2 < \infty$,

$$Z_T^{-1}\Sigma_{\gamma \varepsilon \Gamma_T} 1[T^{-1/2} X_T(\gamma) \varepsilon A] \to B(A) \quad \text{a.s.}$$

as $T \to \infty$, where B and A are as in the last result. Again this is an invariance theorem that extends the respective central limit theorem.

There are strong laws of large numbers related to the last two results. Suppose that to each node θ in the family tree there corresponds a random variable $V(\theta)$, such that the $V(\theta)$ are independent of the tree, independent of each other, with common mean EV and uniformly bounded second moments. Let Θ_i denote the set of nodes in the i-th generation. In the Galton-Watson tree, for $i \leq n$ and $\theta \varepsilon \Theta_i$, let ξ_n^θ denote the number of descendants of θ in the n-th generation. If $\{k_n\}$ are non-decreasing positive integers such that $k_n \leq n$, then

$$k_n^{-1}\Sigma_{i=1}^{k_n} \xi_n^{-1}\Sigma_{\theta \varepsilon \Theta_i} \xi_n^\theta V(\theta) \to EV \quad \text{a.s.}$$

as $n \to \infty$. Similarly, for an age-dependent Markov branching population, let Z_T^θ denote the number of descendants of the node θ that are living at time T (if θ is at a time greater than T, then $Z_T^\theta = 0$), and let $\{k_T\}$ be non-decreasing generation numbers. Then

$$k_T^{-1}\Sigma_{i=1}^{k_T} Z_T^{-1}\Sigma_{\theta \in \Theta_i} Z_T^\theta V(\theta) \to EV \quad \text{a.s.}$$

as $T \to \infty$. These two laws of large numbers follow from martingale arguments in [12, 13].

4. EXTENSIONS

We now consider a more general situation than in the previous section.

Suppose that to each branch $\gamma \in \Gamma_T$ there corresponds a <u>branch process</u> $X_T(\gamma)$, random element of $D[0,1]$, the space of real functions on $[0,1]$ that are right-continuous with left limits, endowed with the Skorohod topology. The value of $X_T(\gamma)$ at time t is interpreted as the position of the offspring on γ alive at time Tt, and the increment process of $X_T(\gamma)$ associated to the life span of an offspring is its corresponding <u>offspring process</u>. Usually, the offspring processes are given and the branch processes are obtained by piecing the offspring processes along the branches. We assume, as is the case in the previous examples, that all the branch processes $X_T(\gamma)$, $\gamma \in \Gamma_T$, have the same distribution (notice that a branch process includes the randomness of the split times on the branch in the age-dependent case).

In the examples of Section 3, the branch processes converge weakly with a normalization as n or $T \to \infty$, and they all have the same limit. For the Galton-Watson branching random walk, the branch processes are defined by the successive partial sums of the offspring jumps along the branches, and the limit is Brownian motion (Donsker's theorem). In the case of a Bellman-Harris branching Brownian motion, the branch processes are Brownian motions, and hence they converge trivially to Brownian motion. For a Bellman-Harris branching random walk, the branch processes are random walks, and they converge weakly to Brownian motion with variance parameter the reciprocal mean of the offspring lifetime distribution (see [16]). Other branch processes may be considered that converge to diffusions different from Brownian motion (e.g. certain transport processes [23]).

In general, then, we associate identically distributed branch processes $X_T(\gamma)$ to the branches $\gamma \varepsilon \Gamma_T$, and we suppose that for all γ, $a_T^{-1}X_T(\gamma)$ converges weakly as $T \to \infty$ to a random element X of $D[0,1]$, where a_T is a normalization that goes to ∞. We refer to this assumption as the <u>convergence hypothesis</u>.

We make other assumptions for technical reasons: Clearly, the offspring lifetimes are the durations of the corresponding offspring processes; but otherwise the branch processes are assumed independent of the family tree. The maximal distances attained by the offspring from their birth positions are independent of each-other, with the same mean, and uniformly bounded second moments. Finally, there is a constant K such that

$$E \sup_{s/T \leq t \leq r/T} |X_T(t,\gamma) - X_T(s/T,\gamma)| \leq K(1+r-s)$$

for all $0 \leq s < r \leq T$, and all γ. Such conditions are satisfied by Brownian motion, random walk, and many transport processes. Observe that these assumptions allow the offspring processes along a branch to be dependent and non-identically distributed; this situation occurs in non-isotropic transport processes.

The <u>empirical distribution of the branch processes</u> at time T is

$$Z_T^{-1}\Sigma_{\gamma \varepsilon \Gamma_T} 1[a_T^{-1}X_T(\gamma) \varepsilon \cdot],$$

defined on the Borel sets of $D[0,1]$. It is a random distribution (it depends on realizations of the family tree and the branch processes). The desired invariance principle is of the following form:

with probability one, the empirical distribution of the branch processes converges weakly as $T \to \infty$ to a distribution on $D[0,1]$ which is in some way related to the limit X of the individual branch processes given in the convergence hypothesis.

An approach for proving such an invariance principle is contained in [13], in the case of an age-dependent Markov branching process. Here we will only mention the general idea of the method. First, if $X_n, n = 1,2,\ldots,$ and X are random elements of $D[0,1]$, and $Ef(X_n) \to Ef(X)$ as $n \to \infty$, for all bounded complex functions f on $D[0,1]$ such that for some constant M (depending on f),

$$|f(x)-f(y)| \leqq M \sup_{0 \leqq t \leqq 1} |x(t)-y(t)|, \quad x,y \in D[0,1],$$

then X_n converges weakly to X as $n \to \infty$. (In [12] we have this for $C[0,1]$. The extension to $D[0,1]$ is easy. Appropriate choices for f yield weak convergence of the finite-dimensional distributions and tightness). For such an f as above, let

$$\phi_T = Z_T^{-1} \Sigma_{\gamma \in \Gamma_T} f(a_T^{-1} X_T(\gamma)),$$

and

$$\psi_T = e^{-\alpha(T-\tau_T)} E[\Sigma_{\gamma \in \Gamma_T^{\gamma_0}} f(a_T^{-1} \tilde{X}_T(\gamma)) | \tau_T],$$

where τ_T is a random time (to be defined shortly) smaller than T, $\Gamma_T^{\gamma_0}$ denotes the set of branches in Γ_T that descend from a fixed arbitrary branch $\gamma_0 \in \Gamma_{\tau_T}$, and $\tilde{X}_T(\gamma)$ is given by the expression

$$\tilde{X}_T(t,\gamma) = \begin{cases} 0 & \text{if } 0 \leqq t \leqq \tau_T/T, \\ X_T(t,\gamma) - X_T(\tau_T/T,\gamma) & \text{if } \tau_T/T \leqq t \leqq 1. \end{cases}$$

In [13] we show that under $E\xi_1^2 < \infty$ and the hypotheses stated above, if ψ_T converges a.s. to a constant Q as $T \to \infty$, then ϕ_T

also converges a.s. to Q as $T \to \infty$. The argument proceeds as follows. Let $\{k_T\}$ be non-decreasing generation numbers that go to ∞ as $T \to \infty$, and let τ_T be the maximum of the split times of the k_T-th generation members that split before time T. As a consequence of a result of Kesten (unpublished) and Kingman [20], $\tau_T/T \to 0$ a.s. if $k_T = o(T)$. Then 1) all that occurs up to generation k_T does not count in the limit; 2) the k_T-th generation members give rise to descendant branching processes with branch processes (the $\tilde{X}_T(\gamma)$) that are close to their respective original branch processes (the $X_T(\gamma)$), and these descendant branching motions corresponding to the k_T-th generation members are independent and identically distributed; and 3) the empirical distribution of the branch processes, i.e. ϕ_T, is shown to behave asymptotically as ψ_T, where ψ_T is obtained from an expression analogous to that of ϕ_T for the descendant branching motions, by substituting the sum and the population size by their expectations. Another matter is to identify the limit Q. If Q is identified as $Q = Ef(Y)$ for a random element Y of $D[0,1]$, then it follows, by the above weak convergence fact, that the empirical distribution of the branch processes converges a.s. to the distribution of Y as $T \to \infty$. A complete answer is obtained by relating Y and X.

The above procedure is completed in [13] in the case of Brownian motion offspring processes, and the corresponding argument is carried through in [12] for general offspring processes in the Galton-Watson case. In both models, Y = X. In the age-dependent case, the problem is working with the expectation that appears in ψ_T for general f and general branch processes; the cause is the presence of members of many generations at a given time. Since Brownian motion on a branch is independent of the split times along the branch, it is obvious that for branching Brownian motion Y should equal X; but for branch processes that do depend on the split times, such as random walks and transport processes, it must not be expected that Y and X coincide.

We give an example: Consider an age-dependent Markov branching random walk, and for simplicity suppose that the offspring jumps have mean 0 and variance 1. Let $f(x) = e^{iux(1)}$, $x \in D[0,1]$, where u is a fixed real number. Denote as before $Y_j(\gamma)$ the jump of the j-th offspring on the branch γ, and let $N(t,\gamma)$ be the generation number of the offspring that lives on γ at time t. Let $a_T = T^{1/2}$. Then

$$\psi_T = e^{-\alpha(T-\tau_T)} E[\Sigma_{\gamma \in \Gamma_T^{\gamma_0}} \exp\{iuT^{-1/2} \Sigma_{j=N(\tau_T,\gamma)+1}^{N(T,\gamma)} Y_j(\gamma)\} | \tau_T]$$

$$= e^{-\alpha(T-\tau_T)} E[\Sigma_{\gamma \in \Gamma_{T-\tau_T}} \exp\{iuT^{-1/2} \Sigma_{j=1}^{N(T-\tau_T,\gamma)} Y_j(\gamma)\} | \tau_T].$$

Therefore, since $\tau_T/T \to 0$ a.s., to show that $\psi_T \to Q$ a.s. is equivalent to proving that

$$e^{-\alpha T} E\Sigma_{\gamma \in \Gamma_T} \exp\{iuT^{-1/2} \Sigma_{j=1}^{N(T,\gamma)} Y_j(\gamma)\} \to Q.$$

Let ζ_T stand for the quantity on the left in the last expression. By conditioning on realizations of the tree, and denoting Υ the characteristic function of the offspring jump distribution, we obtain

$$\zeta_T = e^{-\alpha T} E\Sigma_{\gamma \in \Gamma_T} \Upsilon(uT^{-1/2})^{N(T,\gamma)}$$

$$= Ee^{-\alpha T} Z_T Z_T^{-1} \Sigma_{\gamma \in \Gamma_T} \Upsilon(uT^{-1/2})^{N(T,\gamma)}$$

$$= Ee^{-\alpha T} Z_T \Upsilon(uT^{-1/2})^{N^*(T)},$$

where $N^*(T)$ is the generation number of an offspring chosen at random (uniformly) from those alive at time T. Hence

$$\zeta_T = Ee^{-\alpha T} Z_T (\Upsilon(uT^{-1/2})^T)^{N^*(T)/T}.$$

Now, $e^{-\alpha T} Z_T \to W$ a.s. and in the mean, where W is a random variable with mean 1; $T(uT^{-1/2})^T \to e^{-u^2/2}$, by the central limit theorem; and $N^*(T)/T \to 1/\tilde{\mu}$ in probability, where $\tilde{\mu}$ is the mean of the distribution $\tilde{G}$ given in Section 2 (see [24]); for exponential lifetime distribution with parameter λ, we have $\tilde{\mu}=(\lambda m)^{-1}$. Therefore,

$$(T(uT^{-1/2})^T)^{N^*(T)/T} \to e^{-u^2\lambda m/2} \quad \text{in probability,}$$

and by uniform integrability,

$$\zeta_T \to EW e^{-u^2\lambda m/2} = e^{-u^2\lambda m/2}.$$

Since by the argument of [13] the last result implies that

$$Z_T^{-1}\Sigma_{\gamma\varepsilon\Gamma_T} e^{iuT^{-1/2}X_T(1,\gamma)} \to e^{-u^2\lambda m/2} \quad \text{a.s.,}$$

and u is arbitrary, it follows that the empirical distribution of the branching random walk positions at time T, normalized by $T^{1/2}$, converges a.s. as $T\to\infty$ to the gaussian distribution with mean 0 and variance λm. (Recall that for the individual branch random walks the limit variance is λ). This a.s. central limit theorem agrees with the result from Kaplan's theorem [18], where the branching process is Bellman-Harris with minimum moment condition, and the method is different from ours.

We close with some comments.

1) The results we have given hold for motions in several dimensions. In the Galton-Watson case, the offspring jumps can be taken to be the offspring lifetimes.

2) One should prove in general that (essentially as a consequence of the convergence hypothesis) ψ_T converges a.s. to $Ef(Y)$, and identify the distribution of Y in terms of that of X. For the Markov branching random walk, it seems possible to show weak convergence of the finite-dimensional distributions analogously as was done above for the one-dimensional distribution at $t = 1$, and tightness may possibly be argued from the tightness of the individ-

ual branch random walks. Our method in [13] employs at some points the Markovian property of the branching process; a proof that works for general Bellman-Harris populations may possibly be obtained by combining our approach with the methods of Athreya and Kaplan [5,6]. A refinement of technical importance would be to prove the invariance principles under minimum moment conditions on the offspring production law.

3) On the basis of what we have so far, we conjecture that the theorem for Bellman-Harris populations is that the empirical distribution of the branch processes converges a.s. to the limit distribution of the individual branch processes, but with the offspring lifetime distribution G substituted by $\tilde{G}$ (see Section 2). (This substitution is irrelevant in the case of Brownian motion).

4) For applications, the main condition to be verified is the convergence hypothesis for the individual branch processes. An interesting case is that of branching transport processes, where a population of particles evolves as an age-dependent branching process, and a particle moves, for example, with constant velocity until it splits, and the directions of the offspring particles are distributed with radial symmetry about the direction of the parent particle. For very general transport processes (individual branch), diffusion limits have been obtained by Papanicolaou [23], and for a different case, by Gorostiza [11]. These results yield the convergence hypothesis, and it remains only to prove convergence of ψ_T in order to obtain invariance theorems. It would be desirable also to have invariance theorems for the branching diffusion models contained in the papers cited at the end of Section 2.

5) A possible alternate way of obtaining invariance theorems could be to work first with deterministic trees of a certain type, and then proceed as Joffe and Moncayo [17], by showing that the realizations of the underlying branching process are of that type.

6) Our model requires a normalization $a_T \to \infty$. Moncayo [21] has an a.s. central limit theorem for a binary Galton-Watson

population where there is no normalization, but all the random variables attached to the nodes of the tree up to generation n change their distribution as n increases, and the limit empirical distribution is the Poisson. A modification of our argument may possibly yield a corresponding invariance principle.

REFERENCES

[1] Asmussen, S. Some martingale methods in the limit theory of supercritical branching processes, this volume.

[2] Asmussen, S. and Hering, H. Strong limit theorems for general supercritical branching processes with applications to branching diffusions, *Z. Wahrsch. verw. Geb.* 36, 195-212 (1976).

[3] Asmussen, S. and Hering, H. Strong limit theorems for supercritical immigration-branching processes, *Math. Scand.* 39, 327-342 (1976).

[4] Asmussen, S. and Hering, H. Some modified branching diffusion models, *Math. Biosci.* 34 (1977), to appear.

[5] Athreya, K.B. and Kaplan, N. Convergence of the age distribution in the one-dimensional supercritical age-dependent branching process, *Ann. Prob.* 4, No. 1, 38-50 (1976).

[6] Athreya, K.B. and Kaplan, N. The additive property and its applications in branching processes, this volume.

[7] Athreya, K.B. and Ney, P. *Branching Processes*, Springer Verlag, New York, 1973.

[8] Billingsley, P. *Convergence of Probability Measures*, Wiley, New York, 1968.

[9] Conner, H.E. Asymptotic behavior of averaging processes for a branching process of restricted Brownian particles, *J. Math. Anal. Appl.* 20, 464-479 (1967).

[10] Davis, A.W. Branching-diffusion processes with no absorbing boundaries II, *J. Math. Anal. Appl.* 19, 1-25 (1967).

[11] Gorostiza, L.G. Convergence of transport processes with radially symmetric direction changes, and chain molecules, *J.*

Appl. Prob. 12, 812-816 (1975).

[12] Gorostiza, L.G. and Moncayo, A.R. Invariance principle for random processes on Galton-Watson trees, J. Math. Anal. Appl. 60, 461-476 (1977).

[13] Gorostiza, L.G. and Moncayo, A.R. Almost-sure invariance principle for branching Brownian motion, Adv. in Math., to appear.

[14] Harris, T.E. The Theory of Branching Processes, Springer-Verlag, Berlin, 1963.

[15] Hering, H. Minimal moment conditions in the limit theory for general Markov branching processes, Ann. Inst. H. Poincaré Sec. B, 13 (1977), to appear.

[16] Ighehart, D.L. Weak convergence in applied probability, Stoch. Proc. Appl. 2, 211-241 (1974).

[17] Joffe, A. and Moncayo, A.R. Random variables, trees, and branching random walks, Adv. in Math. 10, No. 3, 401-416 (1973).

[18] Kaplan, N. A limit theorem for a branching random walk, to appear.

[19] Kaplan, N. and Asmussen, S. Branching random walks II, Stoch. Proc. Appl. 4, 15-31 (1976).

[20] Kingman, J.F.C. The first birth problem for an age-dependent branching process, Ann. Prob. 3, No. 5, 790-801 (1975).

[21] Moncayo, A.R. Poisson convergence and family trees, Ann. Prob. 3, No. 6, 1059-1061 (1975).

[22] Ney, P.E. The convergence of a random distribution function associated with branching processes, J. Math. Anal. Appl. 12, 316-327 (1965).

[23] Papanicolaou, G.C. Asymptotic analysis of transport processes, Bull. Amer. Math. Soc. 81, No. 2, 330-392 (1975).

[24] Samuels, M.L. Distribution of the branching process population among generations, J. Appl. Prob. 8, 655-667 (1971).

[25] Stam, A.J. On a conjecture by Harris, Z. Wahrsch. verw. Geb. 5, 202-206 (1966).

[26] Watanabe, S. On the branching process for Brownian particles

with an absorbing boundary, J. Math. Kyoto Univ. 4, No. 2, 385-398 (1965).

[27] Watanabe, S. Limit theorems for a class of branching processes, (Markov processes and potential theory), Proc. Sympos. Math. Res. Center, Madison, Wisconsin, Wiley, New York, 205-232 (1967).

MULTIGROUP BRANCHING DIFFUSIONS

Heinrich Hering

Universität Regensburg
and Københavns Universitet

We review some of the basic limit theorems for Markov branching processes in the framework of multigroup branching diffusions on bounded domains with mixed boundary conditions. This setting allows one to exhibit methods of the limit theory for general Markov branching processes without having to impose technical conditions.

1. THE MODEL

Let Ω be the union of K connected open sets Ω_ν, $\nu = 1,\ldots,K$, in an N-dimensional, orientable manifold of class C^∞, let the closures $\bar{\Omega}_\nu$ be compact and pairwise disjoint, and let the boundary $\partial\Omega$ consist of a finite number of simply connected (N-1)-dimensional hypersurfaces of class C^3. Let X be the union of K Borel sets X_ν such that

$$\Omega_\nu \subset X_\nu \subset \overline{\Omega}_\nu, \quad \nu = 1,\dots,K,$$

in a way to be determined, and suppose to be given a uniformly elliptic differential operator $A|D(A)$, represented in local coordinates on X by

$$A: = \sum_{i,j=1}^{N} \frac{1}{\sqrt{a(x)}} \frac{\partial}{\partial x^i} a^{ij}(x)\sqrt{a(x)} \frac{\partial}{\partial x^j} + \sum_{i=1}^{N} b^i(x) \frac{\partial}{\partial x^i}$$

$$D(A): = \{u|_X: u\in C^2(\overline{\Omega}) \wedge (\alpha u+\beta\frac{\partial u}{\partial n})|_{\partial\Omega} = 0\},$$

where (a^{ij}) and (b^i) are the restrictions to X of a symmetric, second-order, contravariant tensor of class $C^{2,\lambda}(\overline{\Omega})$ and a first-order, contravariant tensor of class $C^{1,\lambda}(\overline{\Omega})$,

$$a: = \det(a^{ij})^{-1},$$

$$0 \leq \alpha,\beta \in C^{2,\lambda}(\partial\Omega), \quad \alpha+\beta \equiv 1,$$

$$\overline{\Omega} \setminus X: = \{\beta=0\}.$$

By $\frac{\partial}{\partial n}$ we denote the exterior normal derivative according to (a^{ij}) at $\partial\Omega$.

Define B as the Banach algebra of all complex-valued, bounded, Borel-measurable functions on X with supremum-norm $\|\cdot\|$, B_+ as the cone of all non-negative functions in B, further

$$C^\ell: = \{u|_X: u \in C^\ell(\overline{\Omega})\},$$

$$C_0^\ell: = \{u|_X: u \in C^\ell(\overline{\Omega}) \wedge u|_{\overline{\Omega}\setminus X} \equiv 0\}.$$

As is wellknown, the closure of $A|\{\xi\in D(A): A\xi\in C_0^0\}$ in B is the C_0^0-generator of a contraction semigroup $\{T_t\}_{t\in\mathbb{R}_+}$ in B, which is non-negative respective B_+, stochastically continuous in $t\geq 0$ on B, and strongly continuous in $t\geq 0$ on C_0^0, with $T_t B \subseteq C_o^2$ for $t>0$. This semigroup determines a conservative, continuous, strong Markov process $\{x_t, P^x\}$ on $X \cup \{\partial\}$, where ∂ is a trap.

Suppose to be given a $k \in B_+$, and define $\overline{k}(x) := k(x)$ for $x \in X$, $\overline{k}(\partial) := 0$, and

$$\delta_t := \exp\{-\int_0^t \overline{k}(x_s)\,ds\}.$$

Let $\{x_t^0, P_0^x\}$ be the δ_t-subprocess of $\{x_t, P^x\}$, defined as a conservative process on $X \cup \{\partial\} \cup \{\Delta\}$, where Δ is a trap corresponding to the stopping by δ_t. For $\xi \in B$ define $\xi_0(x) := \xi(x)$, if $x \in X$, and $\xi_0(\partial) := \xi_0(\Delta) := 0$.

Then

$$T_t^0 \xi(x) := E_0^x \xi_0(x_t^0), \quad x \in X, \ t \geq 0,$$

defines a non-negative contraction semigroup $\{T_t^0\}_{t \in \mathbb{R}_+}$ on B. It is the unique solution of

$$(1.1) \quad T_t^0 = T_t - \int_0^t T_s k T_{t-s}^0 \, ds, \quad t \geq 0,$$

and it is stochastically continuous in $t \geq 0$ on B and strongly continuous in $t \geq 0$ on C_0^0, with $T_t^0 B \subseteq C_0^1$ for $t \geq 0$.

Let $X^{(n)}, n \geq 1$, be the symmetrization of the direct product of n disjoint copies of X, $X^{(0)} := \{\theta\}$ with some extra point θ. Define

$$\hat{X} := \bigcup_{n=0}^{\infty} X^{(n)},$$

and let $\hat{A}$ be the σ-algebra on $\hat{X}$ induced by the Borel algebra on X. Define

$$\hat{x}[\xi] := 0, \qquad \hat{x} = \theta,$$

$$:= \sum_{\nu=1}^{n} \xi(x_\nu), \hat{x} = \langle x_1, \ldots, x_n \rangle \in X^{(n)}, \ n > 0$$

for every finite-valued Borel-measurable ξ on X. Suppose to be given a stochastic kernel $\pi | X \otimes \hat{A}$ such that

$$m\xi(x) := \int_{\hat{X}} \hat{x}[\xi]\pi(x,d\hat{x}), \quad \xi \in B, \ x \in X,$$

defines a bounded operator m on B and the K×K-matrix with elements

$$m_{\nu\mu} := \int_{X_\nu} k(x) m 1_{X_\mu}(x)\,dx, \quad \nu,\mu = 1,\ldots,K,$$

is irreducible.

More explicitly, we assume that either

$$m\xi(x) = \int_X m(x,y)\xi(y)\,dy, \quad \xi \in B, \ x \in X,$$

$$m(x,y) \le \bar{m}(x,y), \quad (x,y) \in X \otimes X,$$

$$\bar{m} \in C^1(\bar{\Omega}\otimes\bar{\Omega}), \quad \bar{m}(\cdot,x) = \bar{m}(x,\cdot) \equiv 0, \quad x \in \bar{\Omega}\setminus X,$$

$$dy := \sqrt{a(y)}\,dy^1 \ldots dy^N, \tag{1.2}$$

where $y^1,\ldots,y^N$ are local coordinates of y, or that the connected components $X_\nu, \nu = 1,\ldots,K$, of X are congruent and

$$\pi(x,\hat{A}) = p_{0..0}(x) 1_{\hat{A}}(\theta) + \sum_{\substack{n_1 \ge 0,\ldots,n_K \ge 0 \\ n_1+\ldots+n_K > 0}} p_{n_1..n_K}(x)$$

$$\times 1_{\hat{A}}(\overbrace{\kappa_1 x,\ldots,\kappa_1 x}^{n_1},\ldots,\overbrace{\kappa_K x,\ldots,\kappa_K x}^{n_K}), \quad x \in X, \ \hat{A} \in \hat{\mathcal{A}}$$

where $1_{\hat{A}}$ is the indicator function of $\hat{A}$, $\{p_{n_1..n_K}(x)\}$ a probability distribution on $\mathbb{Z}_+^K$ for every $x \in X$, and $\kappa_\nu x$ the picture of x produced in X_ν by the given congruence. In the second case

$$m\xi(x) = \sum_{\nu=1}^{K} m_\nu(x)\xi(\kappa_\nu x), \quad \xi \in B, x \in X,$$

$$m_\nu := \sum_{n_1 \ge 0,\ldots,n_K \ge 0} n_\nu p_{n_1..n_K}, \quad \nu = 1,\ldots,K.$$

The pair (x_t^0,π) determines a conservative, right-continuous strong Markov process $\{\hat{x}_t, P^{\hat{x}}\}$ on $(\hat{X},\hat{\mathcal{A}})$, constructed according to the following intuitive rules: All particles at a time move independently of each other, each according to $\{x_t^0, P_0^x\}$. A particle hitting ∂ disappears, a particle hitting Δ is replaced by a population of new particles according to $\pi(x_{t_\Delta -},\cdot)$, where $x_{t_\Delta -}$ is the left limit of the path at the hitting time of Δ, cf.[10], [17].

A simple example for the first kind of branching law we have admitted is the following: A branching event at x results with probability $p_{n_1..n_K}(x)$ in $n_1+..+n_K$ new particles, n_ν of them in X_ν, $\nu = 1,..,K$. The places of birth are distributed independently, a location in X_ν with the distribution density $f_\nu(x,\cdot)$, $\nu = 1,..,K$. Then

$$m(x,y) = \sum_{\nu=1}^{K} 1_{X_\nu}(y) f_\nu(x,y) \sum_{n_1,..,n_K \geq 0} n_\nu p_{n_1,..,n_K}(x).$$

The idea behind the second type of branching law is this: There are K different kinds of particles moving on the same physical domain. To the kind ν we assign X_ν as abstract domain of diffusion, $\nu = 1,..,K$. In the physical domain new particles are always born at the termination point (left limit) of their immediate ancestor.

In terms of the generating functional

$$F_t(\hat{x},\eta) := E^{\hat{x}}\tilde{\eta}(\hat{x}_t),$$

$$\tilde{\eta}(\hat{x}) := 1, \qquad \hat{x} = \theta,$$

$$:= \prod_{\nu=1}^{n} \eta(x_\nu), \quad \hat{x} = \langle x_1,..,x_n\rangle,$$

$$t \geq 0, \quad \hat{x}\in\hat{X}, \quad \eta\in\bar{S} := \{\xi\in B: \|\xi\| \leq 1\},$$

the assumption of independent motion and branching takes the form

$$\text{(F.1)} \quad F_t(\hat{x},\eta) = 1, \qquad \hat{x} = \theta,$$

$$= \prod_{\nu=1}^{n} F_t(\langle x_\nu\rangle,\eta), \quad \hat{x} = \langle x_1,..,x_n\rangle, \quad n>0.$$

Defining F_t: $\bar{S}\to\bar{S}$ by $F_t[\cdot](x) := F_t(\langle x\rangle,\cdot)$, $x\in X$, (F.1) combined with the Chapman-Kolmogorov equation yields

$$\text{(F.2)} \quad F_{t+s}[\cdot] = F_t[F_s[\cdot]], \quad t,s \geq 0.$$

For every $t>0$ define $\hat{x}_{t-}$ on $\hat{Y}$ with $Y: = X\cup\{\partial\}$, and let A_0 be the set of open spheres intersected with X. Define

$$\tau: = \inf\{t>0: \exists U\in A_0: \hat{x}_{t-}[1_U]\neq\hat{x}_t[1_U]\}.$$

It follows from the strong Markov property of $\{\hat{x}_t, P^{\hat{x}}\}$ that for every $\eta\in\overline{S}$ the function $F_t[\eta](x)$, $t\geq 0$, $x\in X$, solves

$$\text{(IF)}\qquad u_t(x) = E^{\langle x\rangle}\tilde{\eta}(\hat{x}_t)1_{\{t<\tau\}} + E^{\langle x\rangle}(E^{\hat{x}_\tau}\tilde{\eta}(\hat{x}_{t-s}))|_{s=\tau}1_{\{\tau<s\}}$$

$$= T_t^0\ \eta(x) + P_0^x(x_\tau^0=\partial, \tau\leq t)$$

$$+ \int_0^t\int_X P_0^x(x_\tau^0=\Delta,\ x_{\tau-}^0\in dy,\ \tau\in ds)\int_{\hat{X}}\pi(y,d\hat{x})F_{t-s}(\hat{x},\eta)$$

$$= T_t^0\eta(x) + H_t(x) + \int_0^t T_s^0\{kf[u_{t-s}]\}(x)ds,$$

$$H_t(x): = 1 - T_t^0(x) - \int_0^t T_s^0 k(x)ds,$$

$$f[\eta](x): = \int_{\hat{X}}\pi(x,d\hat{x})\tilde{\eta}(\hat{x}).$$

The uniqueness of the solution is easily verified by use of

$$\|f[\eta] - f[\xi]\| \leq \|m\|\,\|\eta-\xi\|.$$

The assumptions guarantee that for every $t\geq 0$

$$M_t\xi(x): = E^{\langle x\rangle}\hat{x}_t[\xi],\quad \xi\in B,\ x\in X,$$

defines a non-negative, linear-bounded operator M_t on B. It follows from (F.1) that

$$\text{(1.3)}\qquad E^{\hat{x}}\xi(\hat{x}_t) = \hat{x}[M_t\xi],\quad \hat{x}\in\hat{X},\ \xi\in B,\ t\geq 0,$$

and from (F.2) that $\{M_t\}_{t\in\mathbb{R}_+}$ is a semigroup: Simply set $\eta = \zeta+\lambda\xi$, differentiate with respect to λ at $\lambda=0$ and let $\zeta\to 1$, using dominated convergence. Similarly, (IF) implies that for every $\xi\in B$ the function $M_t\xi(x)$, $t\geq 0$, $x\in X$, solves

(IM) $\quad v_t(x) = T_t^0\xi(x) + \int_0^t T_s^0\{kmv_{t-s}\}(x)ds.$

Again, the solution is unique.

Throughout this paper $c_\nu > 0$, $\nu \in \mathbb{N}$, will be suitable real constants.

2. POSITIVITY THEOREM

To obtain a satisfactory limit theory we need a positivity result which is stronger than the conventional Kreĭn-Rutman theorem. Define

$$D_0^+ := \{u|_X : u \in C^1(\overline{\Omega}), u>0 \text{ on } X, u=0 \wedge \frac{\partial u}{\partial n}<0 \text{ on } \overline{\Omega}\setminus X\}.$$

Theorem 1 ([6],[7]). The moment semigroup $\{M_t\}_{t\geq 0}$ is stochastically continuous in $t\geq 0$ on B, strongly continuous in $t\geq 0$ on C_0^0 with $M_t B \subseteq C_0^1$ for $t>0$. It can be represented in the form

(M) $\quad M_t = \rho^t \varphi\Phi^* + \Delta_t, \quad t>0,$

$\quad \Phi^*[\xi] \doteq \int_X \varphi^*(x)\xi(x)dx, \quad \xi \in B,$

where $\quad 0<\rho \in \mathbb{R}, \quad \varphi \in D_0^+, \quad \varphi^* \in D_0^+, \quad \Phi^*[\varphi] = 1,$

and $\quad \Delta_t$: $B \to B$ such that for all $t>0$

$\quad \varphi\Phi^*\Delta_t = \Delta_t\varphi\Phi^* = 0,$

$\quad -\alpha_t\varphi\Phi^* \leq \Delta_t \leq \alpha_t\varphi\Phi^* \quad [B_+],$

with $\alpha.: \mathbb{R}_+ \to \mathbb{R}_+$ satisfying $\rho^{-t}\alpha_t \downarrow 0$ as $t\uparrow\infty$.

Remark. Notice that there are no continuity requirements for k, m(x,y), or $m_\nu(x)$. We do not consider $\{M_t\}$ as extension of a semigroup generated on C_0^0, but as restriction of a semigroup generated on L^2.

Proof. We modify the procedure of [7]. Define $L^2 := L^2(X)$ respective (1.2). Let $\overline{T}_t$ be the extension of T_t to L^2 and $\overline{T}_t^*$ the adjoint of $\overline{T}_t$. Then

$$T_t\xi(x) = \int_X p_t(x,y)\xi(y)dy \quad [B],$$

$$\overline{T}_t^*\xi(x) = \int_X p_t(y,x)\xi(y)dy \quad [L^2],$$

where $p_t(x,y)$ is the fundamental solution of $\partial p_t/\partial t = Ap_t$. That is, $p_t(x,y)$ is given as a continuous function on $\{t>0\}\otimes\overline{\Omega}\otimes\overline{\Omega}$, continuously differentiable in x and y for $t>0$, such that for $0<t\leq t_0$, t_0 arbitrary but fixed,

$$\text{(2.1)} \quad \begin{aligned} &p_t(x,y)>0, \quad (x,y)\in X_\nu\otimes X_\nu, \quad \nu=1,..,K, \\ &p_t(x,y)=0, \quad (x,y)\in X_\nu\otimes X_\mu, \quad \nu\neq\mu, \end{aligned}$$

$$\text{(2.2)} \quad p_t(x,\cdot)=p_t(\cdot,x)\equiv 0, \quad x\in\overline{\Omega}\setminus X,$$

$$\text{(2.3)} \quad \begin{aligned} &\frac{\partial p_t}{\partial n_x}(x,y)<0, \quad (x,y)\in(\overline{\Omega}_\nu\setminus X_\nu)\otimes X_\nu, \\ &\frac{\partial p_t}{\partial n_y}(x,y)<0, \quad (x,y)\in X_\nu\otimes(\overline{\Omega}_\nu\setminus X_\nu), \quad \nu=1,..,K, \end{aligned}$$

$$\text{(2.4)} \quad \sup_{x,y\in X}\{|\frac{\partial p_t}{\partial x^i}(x,y)|+|\frac{\partial p_t}{\partial y^i}(x,y)|\} = O(t^{-(N+1)/2}), \quad i=1,..,N,$$

$$\text{(2.5)} \quad \sup_{x\in X}\int_X\{|\frac{\partial p_t}{\partial x^i}(x,y)|+|\frac{\partial p_t}{\partial x^i}(y,x)|\}dy = O(t^{-1/2}), \quad i=1,..,N,$$

$$\text{(2.6)} \quad p_t(x,y) = c_1 t^{-N/2} \sum_j 1_{Y_j}(x)\exp\{-c_2 t^{-1}\sum_{i=1}^N |x_j^i-y_j^i|^2\} \text{ on } X\otimes X,$$

where $\{Y_j\}$ is a finite covering of $\overline{\Omega}$ by canonical coordinate neighbourhoods and $x_j^1,..,x_j^N$ are the coordinates of x in the coordinate system associated with Y_j, cf. [11], [16]. As an immediate consequence,

$$T_t^+\xi(x) := \int_X p_t(x,y)\xi(y)dy \quad [B]$$

defines a restriction of $\overline{T}_t^*$ to B, whose norm $\|T_t^+\|$ is bounded on bounded t-intervals.

Let $\overline{m}$, $\overline{b}$, and $\overline{T}_t^0$ be the extensions to L^2 of m,

$$b: = km + \|k\| - k$$

and T_t^0. The closure of A+k(m-1) in L^2 generates a semigroup $\overline{M}_t$, which is the unique solution of

$$\overline{M}_t = \overline{T}_t^0 + \int_0^t \overline{T}_s^0 \, k\overline{m}\overline{M}_{t-s} ds \qquad [L^2]$$

and thus is identical with the extension of M_t to L^2. Similary the semigroup

$$\overline{N}_t: = e^{\|k\|t}\overline{M}_t,$$

generated by the closure of A+b in L^2, is the extension to L^2 of the unique solution N_t of

$$(2.7) \quad N_t = T_t + \int_0^t T_s b N_{t-s} ds \qquad [B].$$

Since $T_t B \subseteq C_0^0$, t>0, it follows from (1.1),(IM), and (2.7) by use of $\|T_t\| \leq 1$, the boundedness of k and m, and dominated convergence that $T_t^0 B \subseteq C_0^0$, $M_t B \subseteq C_0^0$, and $N_t B \subseteq C_0^0$ for t>0. Hence also

$$N_t = e^{\|k\|t} M_t.$$

From (2.7)

$$(2.8) \quad N_t = \sum_{k=0}^{\infty} T_t^{(n)}, \quad t \geq 0, \qquad [B]$$

$$T_t^{(0)}: = T_t, \quad T_t^{(n+1)}: = \int_0^t T_s b T_{t-s}^{(n)} ds, \quad n \geq 0 \qquad [B].$$

From this, in particular,

$$(2.9) \quad \|N_t\| \leq e^{\|b\|t}, \quad t \geq 0.$$

Given (2.7) with the bounds for T_t, b, and N_t, strong continuity of $T_t|C_0^0$ implies strong continuity of $N_t|C_0^0$ in $t \geq 0$, and recalling

$N_tB\subseteq C_0^0$, $t>0$, stochastic continuity of $T_t|B$ implies stochastic continuity of $N_t|B$ in $t\geq 0$. Using $p_t(\cdot,y)\in C^1$, (2.4), (2.5), and dominated convergence, we also get $N_tB\subseteq C^1$ for $t\geq 0$.

By continuity of $p_t(x,y)$ in (x,y) with (2.2), $T_t|B$ and $T_t|C_0^0$ are compact, if $t>0$. For $0<\varepsilon<t$ rewrite (2.7) as

$$N_t = T_t+T_\varepsilon\int_\varepsilon^t T_{s-\varepsilon}bN_{t-s}ds+\int_0^\varepsilon T_s bN_{t-s}ds$$

and note that the integrals on the right are bounded operators, the norm of the second being $O(\varepsilon)$ if t is fixed. That is, compactness of T_t implies compactness of N_t for $t>0$. The cone B_+ and its dual B_+^* are closed, have a non-empty interior, and span B and its dual B^*. By (2.1) the spectral radius of T_t and thus the spectral radius σ_t of N_t are positive. Hence, the spectrum of N_t is purely discrete, each non-zero eigenvalue has finite multiplicity, and there exist non-trivial $\varphi_t\in B_+$, $\Phi_t^*\in B_+^*$ such that

$$N_t\varphi_t = \sigma_t\varphi_t, \quad \Phi_t^*N_t = \sigma_t\Phi_t^*,$$

cf.[15]. The same holds for $N_t|C_0^0$. Since $T_sB\subseteq C_0^0$, $s>0$, the spectral radius is the same, and we can take the same $\varphi_t\in C_0^0\cap B_+$ in both cases. For $\varepsilon>0$, $n\leq \ell$, $t>(\ell+1)\varepsilon$ define

$$T_t^{(0,\varepsilon)} := T_{t-\varepsilon}, \quad T_t^{(n,\varepsilon)} := \int_\varepsilon^{t-n\varepsilon} T_{s-\varepsilon}bT_\varepsilon T_{t-s}^{(n-1,\varepsilon)}ds.$$

By (2.8)

$$N_t \geq T_\varepsilon \sum_{n=0}^{\ell} T_t^{(n,\varepsilon)} \qquad [B_+].$$

Fixing t and choosing any $\xi\in B_+$ which is positive on a set of positive measure, we can by (2.1) and the irreducibility assumption on $(m_{\nu\mu})$ find $\varepsilon>0$ and $\ell\in\mathbb{N}$ such that $t>(\ell+1)\varepsilon$ and

$$\sum_{n=0}^{\ell} T_t^{(n,\varepsilon)} \xi \geq c_3 > 0 \text{ on } \Omega' = \cup_{\nu=1}^{K} \Omega'_\nu,$$

where all the $\Omega'_\nu \subset X_\nu$, $\nu=1,..,K$, have positive measure. By (2.1), (2.3), (2.4), and l'Hospital's rule

$$\sup_{x \in X} (T_\varepsilon 1_{X \setminus \Omega'}(x)/T_\varepsilon 1_{\Omega'}(x)) < \infty.$$

That is, there exists a $\delta \in \mathbb{R}_+$, $\delta > 0$, such that $T_\varepsilon 1_{\Omega'} \geq \delta T_\varepsilon 1_X$ and hence

$$N_t \xi \geq c_3 \delta T_\varepsilon 1_X.$$

On the other hand $T_\varepsilon 1_X \in D_0^+$, so that by (2.4), (2,5), (2,7), (2.9, and l'Hospital's rule there exist a $c_4 > 0$ such that

$$N_t \xi \leq c_4 T_\varepsilon 1_X.$$

Consequently, σ_t is a simple eigenvalue of $N_t | C_0^0$, larger in absolute value that any other eigenvalue, cf.[14, Chapter 2]. Once again referring to $N_t B \subseteq C_0^0$, $t>0$, the same is true for $N_t | B$. The $(T_\varepsilon 1_X)$ - boundedness from below, $T_\varepsilon 1_X \in D_0^+$, and $N_t B \subseteq C_0^1$ imply $\varphi_t \in D_0^+$ and thus $\Phi_s^*[\varphi_t] > 0$ for $s,t>0$. Using the semigroup property of N_t, it follows that $\sigma_r =: \sigma^r$ for all rational r and, since σ^r is simple, that $\varphi_r =: \varphi$ for all rational r. By continuity of N_t therefore $\sigma_t = \sigma^t$ and $\varphi_t = \varphi$ for all $t>0$.

Now consider the problem in L^2. Again, $\overline{N}_t$ is compact, the spectral radius $\overline{\sigma}_t$ of $\overline{N}_t$ is in the spectrum of $\overline{N}_t$, and there exist non-trivial, non-negative $\overline{\varphi}_t, \overline{\varphi}_t^* \in L^2$ such that

$$\overline{N}_t \overline{\varphi}_t = \overline{\sigma}_t \overline{\varphi}_t, \quad \overline{N}_t^* \overline{\varphi}_t^* = \overline{\sigma}_t \varphi_t^*,$$

where $\overline{N}_t^*$ is the adjoint of $\overline{N}_t$. However,

$$\overline{\sigma}_t \int_X \overline{\varphi}_t^*(x) \varphi(x) dx = \int_X \overline{N}_t^* \overline{\varphi}_t^*(x) \varphi(x) dx$$

$$= \int_X \overline{\varphi}_t^*(x) N_t \varphi(x) dx = \sigma^t \int_X \overline{\varphi}_t^*(x) \varphi(x) dx > 0.$$

That is, $\bar{\sigma}_t \equiv \sigma^t$, and we can take $\bar{\varphi}_t \equiv \varphi$. Viewing $\bar{T}_t$ as the perturbed semigroup with $\bar{N}_t$ as the unperturbed semigroup and adjoining the corresponding perturbation equation, or simply considering $\bar{N}_t^*$ as generated by the adjoint of $A+\bar{b}$, we get

$$\bar{N}_t^* = \bar{T}_t^* + \int_0^t \bar{T}_s^* \bar{b}^* \bar{N}_{t-s}^* ds \qquad [L^2].$$

Since $\|T_t^+\|$ is bounded on bounded t-intervals and $\bar{b}^*$ has a bounded restriction b^+ to B by assumption, the unique solution of

$$N_t^+ = T_t^+ + \int_0^t T_s^+ b^+ N_{t-s}^+ ds \qquad [B]$$

is a restriction of $\bar{N}_t^*$ to B. It can be written in the form

$$N_t^+ = \sum_{n=0}^{\infty} T_t^{+(0)}, \quad T_t^{+(0)} := T_t^+, \quad T_t^{+(n)} = \int_0^t T_s^+ b^+ N_{t-s}^{+(n-1)} ds,$$

which implies, in particular, that $\|N_t^+\|$ is bounded on bounded t-intervals. We can now repeat for N_t^+ the argument used for N_t and obtain

$$\Phi_t^*[\xi] \equiv \Phi^*[\xi] = \int_X \varphi^*(x)\xi(x)dx, \quad \xi \in B,$$

with $\varphi^* \in D_0^+$. We normalize $\Phi^*[\varphi]=1$.

Summing up, we have shown that

$$N_t = \sigma^t \varphi \Phi^* + \Gamma_t, \quad t>0,$$

with $\Gamma_t : B \to B$ such that

$$\varphi \Phi^* \Gamma_t = \Gamma_t \varphi \Phi^* = 0, \quad t>0,$$

$$\|\Gamma_{n\varepsilon}\| = 0(\vartheta_\varepsilon^n), \quad n \in \mathbb{N},$$

where $\vartheta_\varepsilon \in \mathbb{R}_+$, $\vartheta_\varepsilon < \sigma^\varepsilon$, for every $\varepsilon>0$. Since $\{\Gamma_t\}$ is a semigroup

$$\|\Gamma_t\| \le \sup_{0 \le s \le \varepsilon} \|\Gamma_s\| \; \|\Gamma_{[t/\varepsilon]\varepsilon}\|$$

$$\le \left(\sup_{0 \le s \le \varepsilon} \|N_s\| + \max\{1, \sigma^\varepsilon\} \|\varphi\| \; \Phi^*[1] \right) \|\Gamma_{[t/\varepsilon]\varepsilon}\|, \quad t>0.$$

That is,

$$\|\Gamma_t\| = O(\vartheta^t), \quad t>0,$$

with some $\vartheta \in \mathbb{R}_+, \vartheta<\sigma$. If $0<2\varepsilon<t$,

$$|\Gamma_t \xi(x)| \leq \sup_{\|\eta\|=1} |\Gamma_\varepsilon \eta(x)| \, \|\Gamma_{t-2\varepsilon}\| \, \|\Gamma_\varepsilon \xi\|$$

$$\leq (N_\varepsilon 1(x)+\sigma^\varepsilon \varphi(x)\Phi^*[1])\|\Gamma_{t-2\varepsilon}\|$$

$$\times \quad (\|N_\varepsilon \xi\| + \sigma^\varepsilon \|\varphi\| \, \Phi^*[|\xi|]).$$

Since $N_\varepsilon 1 \in D_0^+$, as well as $\varphi \in D_0^+$, we have $N_\varepsilon 1 \leq C_\varepsilon \varphi$, where C_ε depends only on ε. As

$$|\int_Y N_\varepsilon \xi(x)dx| \leq \int_X |\xi(y)| N_\varepsilon^+ 1_Y(y)dy, \quad Y \in A,$$

all that is left to be shown is

(2.10) $N_\varepsilon^+ 1_Y \leq C_\varepsilon^* \varphi^* \int_Y dx, \quad Y \in A,$

with C_ε^* depending only on ε.

Using (2.4) and $\varphi^* \in D_0^+$,

$$T_\varepsilon^{+(0)} 1_Y \leq c_5 \varepsilon^{-(N+1)/2} \varphi^* \int_Y dx, \quad Y \in A.$$

For $n \geq 1$

$$T_t^{+(0)} = \int_0^t T_s^+ \tilde{T}_{t-s}^{(n)} ds, \quad \tilde{T}_t^{(1)} := b^+ T_t^+, \quad \tilde{T}_t^{(n+1)} := b^+ \int_0^t T_s^+ \tilde{T}_{t-s}^{(n)} ds.$$

By (2.6) we have $p_t(x,y) \leq c_6 \tilde{p}_t(x,y)$ with

$$\int_X \tilde{p}_s(x,z)\tilde{p}_{t-s}(z,y)dz \leq \tilde{p}_t(x,y), \quad 0 \leq s \leq t,$$

and, if the X_ν are congruent,

$$\tilde{p}_t(\kappa_\nu x, \kappa_\nu y) = \tilde{p}_t(x,y), \quad \nu=1,..,K.$$

Using these properties, it is verified by induction that

(2.11) $\tilde{T}_t^{(n)} 1_Y(x) \leq c_7^n \dfrac{t^{n-1}}{(n-1)!} \left(\int_Y dy + \int_Y p_t(x,y)dy \right)$

where $\int_Y dy$ occurs only with the first kind of branching law. From (2.11) by (2.4), (2.5), and $\varphi^* \in D_0^+$

$$T^{+(n)}1_Y(x) = \frac{c_7^n}{(n-1)!}\Big(\int_0^{\varepsilon}\int_X p_s(z,x)(\varepsilon-s)^{n-1}dzds\int_Y dy$$

$$+\{\int_0^{\varepsilon/2}+\int_{\varepsilon/2}^{\varepsilon}\}\int_X p_s(z,x)(\varepsilon-s)^{n-1}\int_Y \tilde{p}_{\varepsilon-s}(z,y)dydzds\Big)$$

$$\leq \frac{c_7^n}{(n-1)!}\Big[c_8\varepsilon^{n-1/2}+c_9\varepsilon^{-(N-1)/2}(\tfrac{\varepsilon}{2})^{n-1}$$

$$+c_{10}\varepsilon^{-(N+1)/2}\tfrac{1}{n}(\tfrac{\varepsilon}{2})^n\Big]\varphi^*(x)\int_Y dz,$$

which implies (2.10).

3. THREE LEMMATA

By first-order Taylor expansion

(FM) $1-F_t[\xi] = M_t[1-\xi]-R_t(\xi)[1-\xi], \quad \xi\in\overline{S},$

$R_t(\eta)\zeta(x) := E^{\langle x\rangle}\omega(\eta,\zeta,\hat{x}_t),$

$\omega(\eta,\zeta,\hat{x}) := 0, \quad \hat{x}[1]\leq 1,$

$$:= \sum_{\nu=1}^{n}\zeta(x_\nu)\Big(1-\int_0^1\prod_{\mu\neq\nu}[1-\lambda(1-\eta(x_\mu))]d\lambda\Big),$$

$$\hat{x} = \langle x_1,\ldots,x_n\rangle,\ n>1.$$

The mapping $R_t(\cdot)[\cdot]:\overline{S}\otimes B\to B$ is sequentially continuous respective the product topology on bounded regions, non-increasing in the first and linear-bounded in the second variable, and it satisfies

(RM) $0=R_t(1)\zeta \leq R_t(\eta)\zeta \leq M_t\eta, \quad (\eta,\zeta)\in\overline{S}_+\otimes B_+,$

where $\overline{S}_+ := \overline{S}\cap B_+$.

Lemma 1. ([5],[8]). For every $t>0$ there exists a mapping $g_t:\overline{S}_+\to B_+$ such that

(R)
$$R_t(\xi)[1-\xi] = g_t[\xi]\rho^t\Phi^*[1-\xi]\varphi, \quad \xi\in\overline{S}_+,$$

$$\lim_{\|1-\xi\|\to 0}\|g_t[\xi]\| = 0,$$

where the convergence is uniform in t on any closed, bounded interval $[a,b]$ with $a>0$.

<u>Proof</u>. It follows from (IF), (IM), (FM), and the corresponding expansion for f,

$$1-f[\xi] = m[1-\xi]-r(\xi)[1-\xi], \quad \xi\in\overline{S},$$

that for every $\varepsilon>0$ and $\xi\in\overline{S}_+$ the function $R_t(\xi)[1-\xi](x)$, $t\geq\varepsilon$, $x\in X$, solves

$$(3.1) \qquad w_t(x) = A_t(x)+B_t^\varepsilon(x)+\int_0^{t-\varepsilon} T_s^0\{kmw_{t-s}\}(x)dx,$$

$$A_t(x) := \int_0^t T_s^0\{kr(F_{t-s}[\xi])[1-F_{t-s}[\xi]]\}(x)ds,$$

$$B_t^\varepsilon(x) := \int_0^\varepsilon T_{t-s}^0\{kmR_s(\xi)[1-\xi]\}(x)ds.$$

As is easily verified, $R_t(\xi)[1-\xi]$ is the only solution bounded in $[\varepsilon,\varepsilon+\lambda]$ for any $\lambda>0$, and it thus equals the limit of the (non-decreasing) iteration sequence $(w_t^{(\nu)}(x))_{\nu\in\mathbb{Z}_+}$, $w_t^{(0)}\equiv 0$.

Suppose $0<\delta<\varepsilon/2$ and $\xi\in\overline{S}_+$. By (FM) and (RM) there exist a $c_{11}>0$ such that for $\delta\leq s\leq t-\delta$ and $t\leq\varepsilon+\lambda$

$$(3.2) \qquad F_{t-s}[\xi] \geq 1-c_{11}\|1-\xi\|.$$

By (IM)

$$(3.3) \qquad T_t^0 \leq M_t \qquad [B_+].$$

Further

$$(3.4) \qquad 0=r(1)\eta \leq r(\zeta)\eta \leq m\eta, \quad (\zeta,\eta)\in S_+\otimes B_+.$$

Finally, recalling the assumptions on m and the fact that $\varphi,\varphi^*\in D_0^+$, there exist constants c and c^* such that

$$(3.5) \qquad km\varphi \leq c\varphi,$$

$$(3.6) \qquad \Phi^*[km\eta] \leq c^*\Phi^*[\eta], \quad \eta\in B_+.$$

Using (3.2-6) and (M),

$$A_t \le \left\{\int_0^\delta + \int_{t-\delta}^t\right\} M_s[kmM_{t-s}[1-\xi]]ds$$

$$+ \int_\delta^{t-\delta} M_s[kr(1-c_{11}\,\|1-\xi\|\,)M_{t-s}[1-\xi]]ds$$

$$\le \delta(c+c^*)(1+\rho^{-\varepsilon/2}\alpha_{\varepsilon/2})\rho^t\Phi^*[1-\xi]\varphi$$

$$+ t(1+\rho^{-\delta}\alpha_\delta)(1+\rho^{-\varepsilon/2}\alpha_{\varepsilon/2})\|k\varphi\|$$

$$\times \Phi^*[r(1-c_{11}\|1-\xi\|)[\varphi]]\rho^t\Phi^*[1-\xi]\varphi.$$

Since

$$\lim_{\|1-\xi\|\to 0} \Phi^*[r(1-c_{11}\|1-\xi\|)\varphi] = 0,$$

we can for every $\varepsilon'>0$ first fix $\delta>0$ such that

$$\left\{\int_0^\delta + \int_{t-\delta}^t\right\}(\cdots)ds \le \tfrac{1}{2}t\varepsilon'\Phi^*[1-\xi]\varphi, \quad \xi\in\overline{S}_+,$$

and then choose a $\delta'>0$ such that

$$\int_\delta^{t-\delta} (\cdots)ds \le \tfrac{1}{2}t\varepsilon'\Phi^*[1-\xi]\varphi, \quad \|1-\xi\| <\delta'.$$

That is,

$$A_t \le t\Theta_{\varepsilon,\lambda}[\xi]\rho^t\Phi^*[1-\xi]\varphi, \quad \varepsilon\le t\le\varepsilon+\lambda,$$

(3.7)

$$\lim_{\|1-\xi\|\to 0} \Theta_{\varepsilon,\lambda}[\xi] = 0, \quad \varepsilon,\lambda>0.$$

Using (RM) and the fact that $T^0_{t-s} \le M_{t-\varepsilon}M_{\varepsilon-s}$ on B_+,

(3.8) $$B^\varepsilon_t(x)\le M_{t-\varepsilon}[\int_0^\varepsilon M_{\varepsilon-s}[kmM_s[1-\xi]](x)ds] =: \overline{B}^\varepsilon_t(x), \quad t>\varepsilon.$$

By (IM), (M), and (3.6)

(3.9) $$\int_0^{t-\varepsilon} T^0_s\{km\overline{B}^\varepsilon_{t-s}\}ds \le \overline{B}^\varepsilon_t \le \varepsilon c^*(1+\rho^{-t+\varepsilon}\alpha_{t-\varepsilon})\rho^t\Phi^*[1-\xi]\varphi.$$

From (3.7-9)

$$\lim_{\nu\to\infty} w_t^{(\nu)} \leq \{e^{ct} t\Theta_{\varepsilon,\lambda}[\xi]+\varepsilon c^*(1+\rho^{-t+\varepsilon}\alpha_{t-\varepsilon})\}\rho^t\Phi^*[1-\xi]\varphi,$$

for $\varepsilon<t\leq\varepsilon+\lambda$. Since $\varepsilon,\lambda>0$ were arbitrary, this implies (R).□

Let $P(\cdot,\cdot)$ be any stochastic kernel on $X\otimes\hat{A}$ such that

$$M\xi(x) := \int_{\hat{X}} \hat{x}[\xi]P(x,d\hat{x}),$$

defines a bounded operator M on B. Let $F[\cdot](x)$ be the generating functional of $P(x,\cdot)$, and as in (FM) expand

$$1-F[\xi]=M[1-\xi]-R(\xi)[1-\xi], \quad \xi\in\overline{S}.$$

Let Ψ^* be a non-negative, linear-bounded functional on B, sequentially continuous with respect to the product topology on bounded regions, and let $\psi\in\overline{S}_+$ be positive on X, possibly with $\inf \psi=0$.

Lemma 2 ([5],[8]). Suppose $\lambda\in(0,1)$. Then

$$(3.10)\quad \sum_{\nu=1}^{\infty} \Psi^*[R(1-\lambda^\nu\psi)\psi] < \infty$$

if and only if

$$(3.11)\quad \Psi^*[\int_{\hat{X}} \hat{x}[\psi]\log\hat{x}[\psi]P(\cdot,d\hat{x})] < \infty.$$

Proof. We have

$$\int_0^\infty \Psi^*[R(1-\lambda^t\psi)\psi]dt-\Psi^*[M\psi]\leq$$
$$\leq \sum_{\nu=1}^{\infty} \Psi^*[R(1-\lambda^\nu\psi)\psi] \leq \int^\infty \Psi^*[R(1-\lambda^t\psi)\psi]dt.$$

Substituting $s=s(\hat{x},t):=-\hat{x}[\log(1-\lambda^t\psi)]/\hat{x}[\psi]$, we get

$$\int_0^\infty \Psi^*[R(1-\lambda^t\psi)\psi]dt$$
$$=\Psi^*[\int_{\hat{X}}\int_0^\infty(\exp\{\hat{x}[\log(1-\lambda^t\psi)]\}-1+\lambda^t\hat{x}[\psi])\lambda^{-t}dtP(\cdot,d\hat{x})$$
$$=\Psi^*[\int_{\hat{X}}\int_0^{s(\hat{x},0)}\{s^{-2}(\exp\{-\hat{x}[\psi]s\}-1+\hat{x}[\psi]s)$$
$$+a(\hat{x},s)\}b(\hat{x},s)dsP(\cdot,d\hat{x})],$$

$$a(\hat{x},s(\hat{x},t)):=s^{-2}(\lambda^t-s)\hat{x}[\psi]=\frac{\hat{x}[\lambda^t\psi]-\hat{x}[|\log(1-\lambda^t\psi)|]}{(\hat{x}[\log(1-\lambda^t\psi)]/\hat{x}[\psi])^2},$$

$$b(\hat{x},s(\hat{x},t)):=-\lambda^{-t}s^2\left(\frac{\partial s}{\partial t}\right)^{-1}=\frac{1}{|\log\lambda|}\;\frac{(\hat{x}[\log(1-\lambda^t\psi)])^2}{\hat{x}[\lambda^t\psi]\hat{x}[\lambda^t\psi/(1-\lambda^t\psi)]},$$

Since $a(\hat{x},s(\hat{x},t))$ and $b(\hat{x},s(\hat{x},t))$ are bounded as functions of $(\hat{x},t)\in\hat{X}\otimes\mathbb{R}_+$, even if $\inf\psi=0$, the substitution $u:=\hat{x}[\psi]s$ leads to the equivalence of (3.1o) and

(3.12) $\Psi^*[\int_{\hat{X}}\hat{x}[\psi]\int_0^{\hat{x}[|\log(1-\psi)|]}u^{-2}(e^{-u}-1+u)du P(\cdot,d\hat{x})]<\infty.$

For all $v>0$

$$0<c_{11}\le[\log(1+v)]^{-1}\int_0^v u^{-2}(e^{-u}-1+u)du\le c_{12}<\infty.$$

Hence (3.12) is equivalent to

$$\Psi^*[\int_{\hat{X}}\hat{x}[\psi]\log(1+\hat{x}[|\log(1-\psi)|])P(\cdot,d\hat{x})]<\infty,$$

which in turn is equivalent to (3.11).□

The independence property (F.1) can also be expressed in the following way. Let F_t be the σ-algebra generated on the sample space by $\{\hat{x}_s;s\le t\}$. For $0\le s\le t$ and every non-negative, A-measurable η

(3.13) $\hat{x}_t[\eta]=\sum_{i=1}^{\hat{x}_s[1]}\hat{x}_t^{s,i}[\eta]$ a.s. $[P^{\hat{x}}]$,

where the $\hat{x}_t^{s,i}$, $i=1,..,\hat{x}_s[1]$, are F_t-measurable and independent conditioned on F_s, and for every $\hat{A}\in\hat{A}$

$$P^{\hat{x}}(\hat{x}_t^{s,i}\in\hat{A}|F_s)=P^{\langle x_i\rangle}(\hat{x}_{t-s}\in\hat{A})\quad\text{a.s. }[P^{\hat{x}}]$$

with $\hat{x}_s^{s,i}=\langle x_i\rangle$. The sample space may not be large enough to allow (3.13) for all $s\le t$. However, we shall need this representation only for fixed s, or for t,s restricted to sets of the form $\{n\delta: n=0,1,2,..\}$, $\delta>0$. In both cases there exist processes equivalent to $\{\hat{x}_t, P^{\hat{x}}\}$ which satisfy (3.13). Hence we can use (3.13) for the process itself without loss of generality.

Lemma 3 ([1]). Let $\chi\colon \mathbb{R}_+ \to \mathbb{R}_+$ be concave with $\chi(0)=0$. Then for every $t>0$

$$(3.14)\quad \Phi^*[E^{\langle\cdot\rangle}\hat{x}_t[\varphi]\chi(\hat{x}_t[\varphi])] < \infty$$

if and only if

$$(3.15)\quad \Phi^*[k\int_{\hat{X}}\hat{x}[\varphi]\chi(\hat{x}[\varphi])\pi(\cdot,d\hat{x})] < \infty.$$

Remark. While $\log x$ does not satisfy the assumptions on χ, (3.14) with

$$\chi(x) = 1_{[0,e)}(x)x/e+1_{[e,\infty)}(x)\log x$$

is equivalent to

$$\Phi^*[E^{\langle\cdot\rangle}\hat{x}_t[\varphi]\log\hat{x}_t[\varphi]] < \infty,$$

and the same applies to (3.15).

Proof. We first assume (3.15). Let $0<\tau_1\le\tau_2\le..$ be the branching times of $\{\hat{x}_t, P^{\hat{x}}\}$, i.e. the times of discontinuities in $\hat{x}_t$ not caused by absorption via ∂. Define

$$\tilde{I}_t(\hat{x}) := E^{\hat{x}}\hat{x}_t[\varphi]\chi(\hat{x}_t[\varphi]), \qquad I_t(x) := \tilde{I}_t(\langle x\rangle),$$

$$\tilde{I}_{t,n}(\hat{x}) := E^{\hat{x}}\hat{x}_t[\varphi]\chi(\hat{x}_t[\varphi])1_{\{\tau_{n+1}>t\}}, I_{t,n}(x) := \tilde{I}_{t,n}(\langle x\rangle).$$

Then

$$(3.16)\quad I_{t,n+1}(x) = \int_0^t T_s^0\{k\int_{\hat{X}}\pi(x,d\hat{x})\tilde{I}_{t-s,n}(\hat{x})\}(x)ds + I_{t,o}(x).$$

Let τ_n^i, $n=1,2,..$, be the branching times of $\{\hat{x}_t^{0,i}, P^{\hat{x}}\}$, $i=1,..,\hat{x}_0[1]$. Then

$$(3.17)\quad \hat{x}_t[\varphi]1_{\{\tau_{n+1}>t\}} \le \sum_{i=1}^{\hat{x}_0[1]} \hat{x}_t^{0,i}[\varphi]1_{\{\tau_{n+1}^i>t\}}$$

If S_r is the sum of r independent, non-negative random variables Z_i, then by use of Jensen's inequality

$$(3.18)\quad ES_r\chi(S_r) \le \sum_{i=1}^{r} \{EZ_i\chi(\sum_{j\neq i} EZ_j)+EZ_i\chi(Z_i)\}$$

$$\le ES_r\chi(ES_r)+\sum_{i=1}^{r} EZ_i\chi(Z_i).$$

Applying this to (3.17), we have for $0\le t\le t_0$, t_0 arbitrary but fixed,

$$(3.19)\quad \tilde{I}_{t,n}(\hat{x}) \le \rho^t\hat{x}[\varphi]\chi(\rho^t\hat{x}[\varphi])+\hat{x}[I_{t,n}],$$

$$\int_{\hat{X}}\pi(x,d\hat{x})\tilde{I}_{t,n}(\hat{x}) \le c_{12}+c_{13}\iota(x)+mI_{t,n}(x)$$

$$\iota(x):=\int_{\hat{X}}\pi(x,d\hat{x})\hat{x}[\varphi]\chi(x[\varphi]).$$

Inserting (3.19) into (3.16) and using (3.3), (3.6), we get

$$\Phi^*[I_{t,n+1}] \le c_{14}+c_{15}\Phi^*[k\iota]+c_{16}t\sup_{0\le s\le t}\Phi^*[I_{t,n}],$$

where $\|I_{t,0}\| = \sup_{x\in X}\varphi(x)\chi(\varphi(x))$ has been absorbed into c_{14}. From this, for $0\le t\le t_0$ with $c_{16}t<1$,

$$\Phi^*[I_t] = \sup_n \sup_{0\le s\le t}\Phi^*[I_{s,n}] < \infty.$$

Applying (3.18) to $Z_i=\hat{x}^{t,i}_{t+s}[\varphi]$, $i=1,..,\hat{x}_t[1]$, $t,s\le t_0$,

$$I_{t+s}(x) = E^{\langle x\rangle}E(\hat{x}_{t+s}[\varphi]\chi(\hat{x}_{t+s}[\varphi])\,|\,F_t)$$

$$\le E^{\langle x\rangle}\{\rho^s\hat{x}_t[\varphi]\chi(\rho^s\hat{x}_t[\varphi])+\hat{x}_t[I_s]\}$$

$$\le c_{17}+c_{18}I_t(x)+c_{19}\Phi^*[I_s]\varphi(x).$$

Thus (3.14) holds for all $t\ge 0$.

Now suppose (3.14) holds for some t. By (1.3) and (M) the process $\{\rho^{-t}\hat{x}_t[\varphi],F_t,P^{\hat{x}}\}$ is a martingale. Since $u\chi(u)$ is convex, this implies

$$(3.2o)\quad \tilde{I}_s(\hat{x}) \le c_{2o}+c_{21}\tilde{I}_t(\hat{x}),\quad 0\le s\le t.$$

We have

$$I_s(x) \geq E^{\langle x\rangle} \hat{x}_s[\varphi]\chi(\hat{x}_s[\varphi])1_{\{\tau_1<s\}}$$

$$= \int_0^s T_u^0\{k\int_{\hat{X}}\pi(\cdot,d\hat{x})\tilde{I}_{s-u}(\hat{x})\}(x)du, \quad s\leq t$$

From (IM) and (3.6)

$$(3.21) \quad \Phi^*[T_s^0\xi] \geq (1-c^*s)\rho^s\Phi^*[\xi], \quad s\geq 0,$$

for every non-negative A-measurable ξ. Hence, for $s\leq 1/c^*$

$$\Phi^*[I_s] \geq c_{22}s\Phi^*[k\iota]-c_{23},$$

which implies (3.15).□

4. THE SUBCRITICAL CASE

Note that $P^{\langle x\rangle}(x_t=\theta)=F_t[0](x)$, $t>0$, $x\in X$. Since $F_t[0]$ is non-decreasing in t by (F.2),

$$q(x):=\lim_{t\to\infty}P^{\langle x\rangle}(\hat{x}_t=\theta), \quad x\in X,$$

exists and satisfies $q=F_t[q]$, $t>0$. From (IF)

$$(4.1) \quad 1-F_t[\xi]=T_t^0(1-\xi)+\int_0^t T_s^0\{k(1-f[F_{t-s}[\xi]])\}ds.$$

If $\xi=1$ a.e., then $F_t[\xi]\equiv 1$, $t>0$. However, if $\xi\in\overline{S}_+$ such that $\xi<1$ on a set of positive measure, it follows from (2.1) and the irreducibility assumption on m by iteration of (4.1) that

$$F_t[\xi](x)<1, \quad x\in X, \ t>0.$$

Theorem 2 ([5],[8]). Suppose $\rho<1$. Then $q=1$, and there exists a constant $\gamma\in\mathbb{R}_+$ such that

$$(4.2) \quad \lim_{t\to\infty}\rho^{-t}P^{\hat{x}}(\hat{x}_t\neq\theta)=\gamma\hat{x}[\varphi]$$

uniformly in $\hat{x}\in X^{(n)}$ for every $n>0$. We have $\gamma>0$ if and only if for some (and thus all) $t>0$

$$(X \log X) \quad \Phi^*[E^{\langle\cdot\rangle}\hat{x}_t[\varphi]\log\hat{x}_t[\varphi]] < \infty.$$

Moreover, there exists a probability measure P on $(\hat{X},\hat{A})$ such that

$$\lim_{t\to\infty} P^{\hat{y}}(\hat{x}_t[1_{A_\nu}]=n_\nu;\nu=1,..,j \mid \hat{x}_t \neq \theta) = P(\hat{x}[1_{A_\nu}]=n_\nu;\nu=1,..,j) \tag{4.3}$$

for each finite, measurable decomposition $\{A_\nu\}_{1\leq\nu\leq j}$ of X and uniformly in $\hat{y}\in X^{(n)}$ for every $n>0$. If $\gamma>0$, then

$$\int_{\hat{X}} \hat{x}[\xi]P(d\hat{x}) = \gamma^{-1}\Phi^*[\xi], \quad \xi\in B. \tag{4.4}$$

If $\gamma=0$, then

$$\int_{\hat{X}} \hat{x}[\xi]P(d\hat{x}) = \infty \tag{4.5}$$

for every $\xi\in B_+$ positive on a set of positive measure.

<u>Remark</u>. By Lemma 3 and the remark following it, (X LOG X) is equivalent to

(x log x) $\Phi^*[k\int_{\hat{X}}\pi(\cdot,d\hat{x})\hat{x}[\varphi]\log\hat{x}[\varphi]] < \infty$,

and this in turn is equivalent to

$$\int_X \xi(x)k(x)\int_{\hat{X}}\pi(x,d\hat{x})\hat{x}[\xi]\log\hat{x}[\xi]dx,$$

where ξ is any continuous, positive function on X which concides near $\overline{\Omega}\setminus X$ with a function in D_0^+.

<u>Proof</u>. From (FM), (RM), and (M) with $\rho<1$

$$\begin{aligned} \|1-F_t[\xi]\| &\leq \|1-F_t[0]\| + \|F_t[|\xi|]-F_t[0]\| \\ &\leq 2\|1-F_t[0]\| \leq \rho^t(1+\rho^{-t}\alpha_t)\Phi^*[1]\|\varphi\| \to 0, \quad t\to\infty, \end{aligned} \tag{4.6}$$

uniformly in $\xi\in\overline{S}$. To continue we need

<u>Lemma 4</u>. Given that $\|1-F_t[0]\|\to 0$, as $t\to\infty$, there exists for every $t>0$ a mapping $h_t:\overline{S}_+\to B$ such that

$$\begin{aligned} &1-F_t[\xi]=(1+h_t[\xi])\Phi^*[1-F_t[\xi]]\varphi, \quad t>0, \ \xi\in\overline{S}_+, \\ &\lim_{t\to\infty}\|h_t[\xi]\| = 0 \text{ uniformly in } \xi\in\overline{S}_+. \end{aligned} \tag{4.7}$$

<u>Proof</u>. If $\xi=1$ a.e., then $F_t[\xi]\equiv 1$, and we may take $h_t[\xi]\equiv 0$. Now suppose $\xi<1$ on a set of positive measure, i.e. $F_t[\xi]<1$ on X. From (F.2) and (FM)

$$1-F_t[\xi]=M_s[1-F_{t-s}[\xi]]-R_s(F_{t-s}[\xi])[1-F_{t-s}[\xi]], t>s>0, \ \xi\in\overline{S}.$$

From this by (M) and (R)

$$(1-\rho^{-s}\alpha_s - \|g_s[F_{t-s}[\xi]]\|)\rho^s\Phi^*[1-F_{t-s}[\xi]]\varphi$$

$$\leq 1-F_t[\xi] \leq (1+\rho^{-s}\alpha_s)\rho^s\Phi^*[1-F_{t-s}[\xi]]\varphi .$$

Combining these inequalities with those obtained by applying Φ^* to them,

$$-\frac{2\rho^{-s}\alpha_s+\|g_s[F_{t-s}[\xi]]\|}{1+\rho^{-s}\alpha_s}\varphi \leq \frac{1-F_t[\xi]}{\Phi^*[1-F_t[\xi]]}-\varphi$$

$$\leq \frac{2\rho^{-s}\alpha_s+\|g_s[F_{t-s}[\xi]]\|}{1-\rho^{-s}\alpha_s-\|g_s[F_{t-s}[\xi]]\|}\varphi$$

for $t\geq t^*(s)$ and $s\geq s^*$ with some $t^*(s)<\infty, s^*<\infty$. Now use $\rho^{-s}\alpha_s\to 0$, $s\to\infty$, (R), and $\|1-F_t[\xi]\| \leq \|1-F_t[0]\|\to 0$, $t\to\infty$.□

<u>Proof of Theorem 2 continued</u>. Using (F.2), (FM), (M), and (RM),

$$(4.8)\quad 0 \leq \rho^{-t-s}\Phi^*[1-F_{t+s}[\xi]]$$

$$= \rho^{-t}\Phi^*[1-F_t[\xi]]-\rho^{-t-s}\Phi^*[R_s(F_t[\xi])[1-F_t[\xi]]$$

$$\leq \rho^{-t}\Phi^*[1-F_t[\xi]] \leq \rho^{-t}\Phi^*[1-F_t[0]].$$

Hence, there exists a non-negative, non-increasing functional $\gamma[\cdot]$ on $\overline{S}_+$ such that

$$(4.9)\quad \rho^{-t}\Phi^*[1-F_t[\xi]]\downarrow\gamma[\xi],\ t\uparrow\infty,\quad \xi\in\overline{S}_+.$$

Combined with (F.1), written in the form

$$(4.10)\quad F_t(\langle x_1,\ldots,x_n\rangle,\xi) = \prod_{\nu=1}^{n}(1-(1-F_t[\xi](x_\nu))),$$

this implies (4.2) with $\gamma:=\gamma[0]$. From (4.8) and (4.7)

$$\rho^{-n}\Phi^*[1-F_n[0]] =$$

$$= \rho^{-1}\Phi^*[1-F_1[0]]\prod_{\nu=1}^{n-1}\{1-\rho^{-1}\Phi^*[R_1(F_\nu[0])[(1+h_\nu[0])\varphi]]\}.$$

That is, $\gamma>0$ if and only if

$$\sum_{\nu=1}^{\infty}\Phi^*[R_1(F_\nu[0])[(1+h_\nu[0])\varphi]] < \infty.$$

If $\gamma>0$, there exists by (4.6) a positive real $\varepsilon<\|\varphi\|^{-1}$ such that $1-F_\nu[0]\geq\varepsilon\rho^\nu\varphi$ for all sufficiently large ν, so that

$$\text{(4.11)}\quad \sum_{\nu=1}^{\infty}\Phi^*[R_1(1-\varepsilon\rho^\nu\varphi)\varphi] < \infty,$$

in view of (RM). On the other hand, if $\gamma=0$, there is for every $\varepsilon>0$ a ν_0 such that $1-F_\nu[0]\leq\varepsilon\rho^\nu\varphi$ for all $\nu\geq\nu_0$, and (4.11) cannot hold. That is, $\gamma>0$ if and only if (4.11) is satisfied for some $\varepsilon<\|\varphi\|^{-1}$. Now recall Lemma 2.

The generating functional of $P^{\hat{x}}(\hat{x}_t\in\cdot\,|\,\hat{x}_t\neq\theta)$ is given by

$$G_t(\hat{x},\xi) := \frac{F_t(\hat{x},\xi)-F_t(\hat{x},0)}{1-F_t(\hat{x},0)} = 1-\frac{1-F_t(\hat{x},\xi)}{1-F_t(\hat{x},0)}.$$

Define $G_t\colon \bar{S}\to\bar{S}$ by $G_t[\cdot](x):=G_t(\langle x\rangle,\cdot)$. If there exists a functional G on $\bar{S}_+$ such that

$$\text{(4.12)}\quad \lim_{t\to\infty}|G_t[\xi](x)-G[\xi]| = 0,\quad \xi\in S_+,\ x\in X,$$

and for every sequence $(\xi_n)_{n\in\mathbb{N}}$ in $\bar{S}_+$ with $\xi_n(x)\to1$, $n\to\infty$, $x\in X$,

$$\text{(4.13)}\quad \lim_{n\to\infty}G[\xi_n] = 1,$$

then G is the restriction to $\bar{S}_+$ of the generating functional of a probability distribution P on $(\hat{X},\hat{A})$, cf. [3],[18], using (4.10)

$$\lim_{t\to\infty}G_t(\hat{x},\xi) = G[\xi],\quad \xi\in\bar{S}_+,$$

and from this, by setting $\xi=\Sigma_\nu 1_{A_\nu}\lambda_\nu$, $|\lambda_\nu|\leq1$, $\nu=1,..,j$, and appealing to the continuity theorem for generating functions, (4.3). Uniformity of (4.12) in x entails the proposed uniformity of (4.3). We now prove the existence of a G satisfying (4.12), (4.13), uniformly in x.

If $\gamma>0$, then (4.12) with $G[\xi]=1-\gamma[\xi]/\gamma$ and uniformity in x follows from (4.9) and Lemma 4, and (4.13) is obtained from

$$0 \le \gamma[\xi_n] \le \rho^{-t}\Phi^*[1-F_t[\xi_n]]\to 0,\ n\to\infty,$$

using dominated convergence. In the following we admit $\gamma=0$.

Lemma 5. For every $t>0$ and $\xi\in\overline{S}_+$ the function $(1-F_t[\xi])/\varphi$ has a continuous extension to $\overline{\Omega}$.

Proof. Recall (4.1). Since all quantities in the integrand are uniformly bounded and $T_s^0 B\subseteq C_0^0$, $s>0$, we have $1-F_t[\xi]\in C_0^0$, $t>0$. That is, $(1-F_t[\xi])/\varphi$ is continuous on X. The continuous extendability to $\overline{\Omega}$ follows by use of $\varphi\in D_+^0$, $T_s^0 B\subseteq C_0^1$, $s>0$, (2.5) and l´Hospital's rule.□

Proof of Theorem 2 continued. Fix $\xi\in\overline{S}_+$. By Lemma 5 the function $h_t[\xi](x)$ of Lemma 4 has a continuous extension $\overline{h}_t[\xi](x)$ to $\overline{\Omega}$ for every $t>0$. Hence, there exists a t_0 such that $G_t[\xi](x)$ has a continuous extension $\overline{G}_t[\xi](x)$ to $\overline{\Omega}$ for every $t\ge t_0$. Since $\overline{\Omega}$ is compact, there must then for each $t\ge t_0$ exist an $\overline{x}_t\in\overline{\Omega}$ such that $\overline{G}_t[\xi](\overline{x}_t)=\|G_t[\xi]\|$. It follows by the same argument as in [12, p.421] that $\overline{G}_t[\xi](\overline{x}_t)$ is decreasing, as $t\to\infty$. Thus

$$(4.14)\quad G[\xi] := \lim_{t\to\infty} \overline{G}_t[\xi](\overline{x}_t)$$

exists. However, for all $t\ge t_0$,

$$1-G_t[\xi]=\frac{1+h_t[\xi]}{1+h_t[0]}\cdot\frac{1+\overline{h}_t[0](\overline{x}_t)}{1+\overline{h}_t[\xi](\overline{x}_t)}\,(1-\overline{G}_t[\xi](\overline{x}_t)),$$

so that (4.14) and Lemma 4 imply (4.12) with uniformity in x.

Using Lemma 4, (F.2), (FM), and (4.1),

$$(4.15)\quad 1-G[F_t[\xi]] = \lim_{s\to\infty}\frac{\Phi^*[1-F_t[F_s[\xi]]]}{\Phi^*[1-F_s[0]]} =$$

$$= \rho^t(1-G[\xi]) - \lim_{s\to\infty}\Phi^*\left[R_t(F_s[0])\ \frac{1-F_s[\xi]}{\phi^*[1-F_s[0]]}\right]$$

$$= \rho^t(1-G[\xi]),\quad t>0,\quad \xi\in\overline{S}_+.$$

In particular,

(4.16) $G[F_t[0]] = 1-\rho^t$.

Now let $(\xi_n)_{n\in\mathbb{N}}$ be any sequence in $\overline{S}_+$ with $\xi_n(x)\to 1$, $n\to\infty$, $x\in X$. Fix $\delta>0$, $s>0$, $n_0>0$ such that

$$\rho^{-\delta}\alpha_\delta<1,$$

$$c_{24}:=\sup_{\xi\in\overline{S}_+:\|1-\xi\|\le\|1-F_s[0]\|}\|g_\delta[\xi]\|<1-\rho^{-\delta}\alpha_\delta,$$

$$(\rho+\alpha_1)\Phi^*[1-\xi_n]\le\rho^\delta(1-\rho^{-\delta}\alpha_\delta-c_{24})\Phi^*[1-F_s[0]],\ n\ge n_0.$$

By (M), (4.1), and (R) this is clearly possible. In view of (4.1), the monotony of $F_t[0]$, (F.2), (FM), (MR), (M), and (R), there exists a sequence of integers $(\ell(n))_{n\in\mathbb{N}}$ such that $\ell(n)\ge s$ if $n\ge n_0$, $\ell(n)\to\infty$ as $n\to\infty$, and

$$\begin{aligned}1-F_1[\xi_n]&\le(\rho-\alpha_1)\Phi^*[1-\xi_n]\varphi\\ &\le\rho^\delta(1-\rho^{-\delta}\alpha_\delta-c_{24})\Phi^*[1-F_{\ell(n)}[0]]\varphi\\ &\le 1-F_{\delta+\ell(n)}[0],\quad n\ge n_0.\end{aligned}$$

Hence, by (4.15), (4.16),

$$\begin{aligned}1\ge G[\xi_n]&=1-\rho^{-1}(1-G[F_1[\xi_n]])\\ &\ge 1-\rho^{-1}(1-G[F_{\delta+\ell(n)}[0]])\\ &=1-\rho^{\delta+\ell(n)-1},\quad n\ge n_0,\end{aligned}$$

which implies (4.13).

To derive (4.4), suppose $\gamma>0$. Then by (4.2)

$$\lim_{t\to\infty}E^{\langle x\rangle}(\hat{x}_t[1]\,|\,\hat{x}_t\neq\theta) = \gamma^{-1}\phi^*[1]<\infty.$$

Hence, G has a bounded first moment functional M. From (4.15)

$$M[M_t\xi] = \rho^t M\xi.$$

By (M) therefore $M=\varepsilon\Phi^*$, ε a positive real number. Using (4.16) and expanding G similarly as F_t in (FM),

$$1 = \rho^{-t}(1-G[F_t[0]])$$

$$= M[\rho^{-t}(1-F_t[0])]-R(F_t[0])[\rho^{-t}(1-F_t[0])],$$

where $R(\zeta)[\xi]$ is linear-bounded in ξ and tends to 0, as $\|1-\zeta\| \to 0$. From this, by (4.2), $1=M[\gamma\varphi]$. That is, $\varepsilon=\gamma^{-1}$.

Now suppose $\gamma=0$, and define

$$\varepsilon_n := \Phi^*[1-F_n[0]]/\Phi^*[1], \quad n\in \mathbb{N}.$$

By (4.1) and the monotony of $F_n[0]$, $0<\varepsilon_n\downarrow 0$, as $n\uparrow\infty$. Fix $t>0$, $n_1>0$, $s>0$ such that

$$\rho^{-t}\alpha_t < 1,$$

$$\rho^{-s}\alpha_s<1, \ (\rho^t-\alpha_t-\rho^t c_{25})/(\rho^s+\alpha_s)\geq 1,$$

$$c_{25} := \sup_{n\geq n_1} \|g_t[1-\varepsilon_n]\|.$$

Due to (M) with $\rho<1$ and (R) this is possible. Then, using (FM), (R), and (M)

$$1-F_t[1-\varepsilon_n]\geq(\rho^t-\alpha_t-\rho^t c_{25})\Phi^*[\varepsilon_n]\varphi\geq 1-F_s[F_n[0]], \quad n\geq n_1.$$

Applying (4.15) and (F.2),

$$(1-G[1-\varepsilon_n])/\varepsilon_n=\rho^{-t}(1-G[F_t[1-\varepsilon_n]])/\varepsilon_n$$

$$\geq\rho^{-t}(1-G[F_s[F_n[0]]])/\varepsilon_n=\rho^{s-t+n}\Phi^*[1]/\Phi^*[1-F_n[0]], \ n\geq n_1.$$

If $\gamma=0$, the last expression tends to ∞, as $n\to\infty$, by (4.2). That is, in this case G cannot have a bounded first moment functional.

5. THE CRITICAL CASE

For t>0 define

$$\mu(t) := \frac{1}{2t}\Phi^*[E^{\langle \cdot \rangle}\{\hat{x}_t[\varphi]^2 - \hat{x}_t[\varphi^2]\}] \le \infty.$$

Proposition. If $\rho=1$, then

$$\mu(t) \equiv \mu := \frac{1}{2}\Phi^*[k\int_{\hat{X}}\pi(\cdot,d\hat{x})\{\hat{x}[\varphi]^2 - \hat{x}[\varphi^2]\}], \quad t>0.$$

Proof. Extend T_t^0, m, M_t, and Φ^* set of all non-negative, not necessarily bounded, A-measurable functions, and define

$$M_t^{(2)}[\xi](x) := E^{\langle x \rangle}\{\hat{x}_t[\xi]^2 - \hat{x}_t[\xi^2]\}, \quad t>0, \ \xi \in B_+, \ x \in X.$$

From (F.2)

$$M_{t+s}^{(2)}[\xi] = M_t^{(2)}[M_s\xi] + M_t M_s^{(2)}[\xi], \quad s,t>0, \ \xi \in B_+$$

Applying (M) with $\rho=1$,

$$(5.1) \quad \Phi^*[M_t^{(2)}[\varphi]] = t\Phi^*[M_1^{(2)}[\varphi]]$$

for all rational t, further

$$\Phi^*[M_t^{(2)}[\varphi]] \ge \Phi^*[M_s^{(2)}[\varphi]], \quad t \ge s.$$

That is, (5.1) holds for all t>0.

By (IF), the function $M_t^{(2)}[\xi](x)$, $t \ge 0$, $x \in X$, finite or not, solves

$$z_t(x) = \int_0^t T_s^0\{kmz_{t-s} + km^{(2)}[M_{t-s}\xi]\}(x)ds,$$

$$m^{(2)}[\xi](x) := \int_{\hat{X}}\pi(x,d\hat{x})\{\hat{x}[\xi]^2 - \hat{x}[\xi^2]\}, \quad \xi \in B_+, \ x \in X.$$

Using (M) with $\rho=1$, (3.3), (3.6), and (3.21),

$$0 \le \Phi^*[\int_0^t T_s^0\{kmM_{t-s}^{(2)}[\varphi]\}ds \le tc^* \sup_{0 \le s \le t} \Phi^*[M_s^{(2)}[\varphi]] = 2c^*t^2\mu,$$

$$t(1-c^*t)\Phi^*[km^{(2)}[\varphi]] \le \Phi^*[\int_0^t T_s^0\{km^{(2)}[\varphi]\}ds] \le t\Phi^*[km^{(2)}[\varphi]], \quad t>0$$

Divide by t and let $t \downarrow 0$.□

Theorem 3 ([4],[8]). Suppose $\rho=1$. Then either $\mu=0$ and

$\hat{x}_t[1]=\hat{x}_0[1]$ a.s. for all $t\geq 0$, or $\mu>0$ and $q\equiv 1$. If $0<\mu<\infty$ then

(5.2) $$\lim_{t\to\infty} tP^{\hat{x}}(\hat{x}_t \neq \theta) = \mu^{-1}\hat{x}[\varphi]$$

uniformly in $\hat{x}\in X^{(n)}$ for each $n>0$, and for every finite, measurable decomposition $\{A_\nu\}_{1\leq\nu\leq j}$ of X and any $\hat{x}\neq\theta$

(5.3) $$\lim_{t\to\infty} P^{\hat{x}}(t^{-1}\hat{x}_t[1_{A_\nu}]\leq\lambda_\nu; \nu=1,..,j\,|\,\hat{x}_t\neq\theta)$$

$$=\begin{cases} 0, & \min_\nu \lambda_\nu \leq 0 \\ 1-\exp\{-\min_\nu[(\mu\Phi^*[1_{A_\nu}])^{-1}\lambda_\nu]\}, & \min_\nu \lambda_\nu > 0 \end{cases}$$

uniformly in $(\lambda_1,..,\lambda_j)\in\mathbb{R}^j$. For $\xi\in B$

(5.4) $$\lim_{t\to\infty} t^{-1}E^{\langle x\rangle}(\hat{x}_t[\xi]\,|\,\hat{x}_t\neq\theta)=\mu\Phi^*[\xi].$$

Remarks. (a) If $\hat{x}_t[1]=\hat{x}_0[1]$ a.s. for all $t>0$, then it follows by (FM) and (M) that φ is constant and, with $\varphi\equiv 1$,

$$\lim_{t\to\infty} P^{\langle x\rangle}(\hat{x}_t[1_A]=1)=\Phi^*[1_A],\ x\in X,\ A\in\mathcal{A}.$$

This case occurs if and only if

$$\int_{\partial\Omega} \alpha(\tilde{y})\,d\tilde{y}+\int_X k(x)\pi(x,\{\hat{x}[1]\neq 1\})\,dx = 0,$$

where $d\tilde{y}$ is the differential surface element of $\partial\Omega$.

(b) As in the case of (x log x) the condition $\mu<\infty$ is equivalent to the condition obtained by substituting for φ and φ^* some continuous positive function which near $\bar{\Omega}\setminus X$ behaves as a function in D_θ^+.

(c) A more intuitive way of expressing (5.3) is the following: The conditional d.f. of the vector $t^{-1}(\hat{x}_t[1_{A_1}],..,\hat{x}_t[1_{A_j}])$, given $\hat{x}_t\neq\theta$, converges to the d.f. of a vector of the form $(\Phi^*[1_{A_1}],..,\Phi^*[1_{A_j}])w$ with $P(w>\lambda)=\exp\{-\lambda/\mu\}$, $\lambda\geq 0$.

Lemma 6. For any finite collection $\{Y_\nu\}_{1\leq\nu\leq j}$ of sets in $\mathcal{A}$

the function $P^{\langle x\rangle}(\hat{x}_t[1_{Y_\nu}]=n_\nu;\nu=1,..,j)$ is continuous in $x\in X$ for every $t>0$ and continuous in $t>0$ for every $x\in X$.

Proof. It suffices to prove the lemma for finite decompositions of X. For any such decomposition

$$P^{\langle x\rangle}(\hat{x}_t[1_{Y_\nu}]=n_\nu;\nu=1,..,j)=H_t(x)+I_t(x), \quad \sum_\nu n_\nu=0$$

$$=\sum_\nu 1_{n_\nu=1}T_t^0 1_{Y_\nu}(x), \quad \sum_\nu n_\nu=1$$

$$=I_t(x), \quad \sum_\nu n_\nu>1$$

$$I_t(x):=\int_0^t T_s^0\{k\int_{\hat{X}}\pi(\cdot,d\hat{x})P^{\hat{x}}(\hat{x}_{t-s}[1_{Y_\nu}]=n_\nu;\nu=1,..,j)\}(x)ds.$$

This follows from (IF). The continuity of $H_t(x)$ and $T_t^0 1_{Y_\nu}(x)$ in x and t and that of $I_t(x)$ in x follows immediately from $\|T_t^0\|\le 1, T_t^0 B\subseteq C_0^0$, $t>0$, and the continuity of T_t^0 in t. As for the continuity of I_t in t, note that

$$\|I_{t+\delta}-I_t\|\le\|T_\delta^0(T_\varepsilon^0 I_{t-\varepsilon})-T_\varepsilon^0 I_{t-\varepsilon}\|+3\|k\|\varepsilon,$$

$$\|I_{t-\delta}-I_t\|\le\|T_{\varepsilon-\delta}^0(T_\varepsilon^0 I_{t-2\varepsilon})-T_\varepsilon^0(T_\varepsilon^0 I_{t-2\varepsilon})\|+4k\|\varepsilon\|,$$

whenever $0<2\delta<2\varepsilon<t$.□

Proof of Theorem 3. Since $\varphi>0$ on X, $\mu=0$ if and only if $\Phi^*[1-P^{\langle\cdot\rangle}(\hat{x}_t\in X^{(1)})]=0$, $t>0$, i.e. $P^{\langle x\rangle}(\hat{x}_t\in X^{(1)})=1$, $x\in X$, $t>0$, by continuity. Now suppose $\mu>0$. Then $P^{\langle x\rangle}(\hat{x}_t\in X^{(1)})\neq 1$ on an x-set of positive measure depending on t. Since by (FM) and (M) with $\rho=1$

$$\Phi^*[P^{\langle\cdot\rangle}(\hat{x}_t=\theta)]=\Phi^*[R_t(0)1],\ t>0,$$

this implies $P^{\langle x\rangle}(\hat{x}_t=\theta)>0$ on a set of positive measure, depending on t. Define

$$N(t):=\{x\in X:P^{\langle x\rangle}(\hat{x}_t=\theta)=0\},\ t>0.$$

Since $P^{\langle x\rangle}(\hat{x}_t=\theta)$ is continuous, N(t) is compact. If $\Phi^*[1_{N(t)}]=0$ for some $t>0$, then $\Phi^*[1-q]=0$ as in [3;|||,12,13].

By $q=F_t[q]$ and Lemma 5, or (FM) and (M) with $\rho=1$, $\Phi^*[1-q]$ implies $q\equiv 1$. Suppose $\Phi^*[1_{N(t)}]>0$ for all $t>0$. Fix s so that $\alpha_s<1$ and define

$$N:=\bigcap_{n\in\mathbb{N}} N(ns).$$

A routine extension of [3;II,6], using compactness of N(t) and thus N and continuity of $P^{\langle x\rangle}(\hat{x}_{2s}[1_N]>1)$ in x, shows that

$$\inf_{x\in N} P^{\langle x\rangle}(\hat{x}_{2s}[1_N]>1)>0$$

and that due to this $\{0<\hat{x}[1]\le d\}$, $0<d<\infty$, is a transient event of $\{\hat{x}_{2ns}, P^{\langle x\rangle}, n\in\mathbb{Z}_+\}$. Given $\rho=1$, this again implies $q\equiv 1$.

<u>Lemma 7</u>. If $\rho=1$ and $\mu<\infty$, then for every $\delta>0$

$$\lim_{\mathbb{N}\ni n\to\infty} \frac{1}{n\delta}\{\Phi^*[1-F_{n\delta}[\xi]]^{-1}-\Phi^*[1-\xi]^{-1}\}=\mu$$

uniformly in $\xi\in\overline{S}_+$ with $\xi<1$ on a set of positive measure.

<u>Proof</u>. Fix ξ as required in the lemma. Then $1-F_t[\xi]>0$ on X for all $t>0$. Using (F.2)

$$\frac{1}{n\delta}\{\Phi^*[1-F_n[\xi]]^{-1}-\Phi^*[1-\xi]^{-1}\}$$

$$=\frac{1}{n}\sum_{\nu=0}^{n-1}\frac{1}{\delta}\{\Phi^*[1-F_\delta[F_{\nu\delta}[\xi]]]^{-1}-\Phi^*[1-F_{\nu\delta}[\xi]]^{-1}\}$$

$$=\frac{1}{n}\sum_{\nu=0}^{n-1}\frac{1}{\delta}(1-\Phi^*[1-F_{\nu\delta}[\xi]]\Lambda_\delta[F_{\nu\delta}[\xi]])^{-1}\Lambda_\delta[F_{\nu\delta}[\xi]],$$

$$\Lambda_\delta[\zeta]:=\Phi^*[1-\zeta]^{-2}\{\Phi^*[1-\zeta]-\Phi^*[1-F_\delta[\zeta]]\}.$$

If $\mu<\infty$, then for $\zeta=1-\eta\varphi\in\overline{S}_+$ with $\eta\in B_+$ and $\xi\in\overline{S}_+$

$$\Phi^*[1-F_t[\zeta]]=\Phi^*[M_t\zeta]-\tfrac{1}{2}\Phi^*[M_t^{(2)}[1-\zeta]]+\tfrac{1}{2}\Phi^*[R_t^{(2)}(\zeta)[1-\zeta]],$$

$$R_t^{(2)}(\xi)[1-\zeta](x):=E^{\langle x\rangle}\omega^{(2)}(\xi,\zeta,\hat{x}_t),$$

$$\omega^{(2)}(\xi,\zeta,\hat{x}):=0,\quad \hat{x}[1]\le 2$$

$$:= \frac{1}{(n-2)!} \sum_{(i_1,..,i_n)} \zeta(x_{\nu_1})\zeta(x_{\nu_2}) \times \left(1-2\int_0^1 (1-\lambda) \prod_{\kappa=3}^{n} [1-\lambda(1-\xi(x_{i_\kappa}))]d\lambda\right),$$

$$\hat{x}=\langle x_1,..,x_n\rangle, \ n>2.$$

By dominated convergence, $\Phi^*[R_t^{(2)}(\cdot)[\cdot]]$ is sequentially continuous on bounded regions in $\overline{S}_+\otimes\{\xi=\eta\varphi:\eta\in B_+\}$, and we have

$$0 = \Phi^*[R_t^{(2)}(1)[\eta\varphi]] \leq \Phi^*[R_t^{(2)}(\xi)[\eta\varphi]]$$
$$\leq \Phi^*[M_t^{(2)}[\eta\varphi]] \leq 2t\mu\|\eta\|^2$$

for $t\geq 0$, $(\xi,\eta)\in \overline{S}_+\otimes B_+$. Using (M) with $\rho=1$ and Lemma 4,

$$\Lambda_\delta[F_t[\xi]]=\tfrac{1}{2}\Phi^*[M_\delta^{(2)}[(1+h_t[\xi])\varphi]]$$
$$-\tfrac{1}{2}\Phi^*[R_\delta^{(2)}(F_t[\xi])[(1+h_t[\xi])\varphi]].$$

Since $1\geq F_t[\xi]\geq F_t[0]\uparrow 1$, as $t\uparrow\infty$,

$$\lim_{t\to\infty} \Lambda_\delta[F_t[\xi]]=\delta\mu$$

uniformly in ξ.□

Proof of Theorem 3 continued. Lemma 7, Lemma 4, and (F.1) written in the form (4.1o) yield (5.2) with t restricted to sets of the form $\{n\delta; n\in \mathbb{N}\}$, $\delta>0$. Since $P^{\hat{x}}(\hat{x}_t=\theta)$ is monotone in t, this implies (5.2) with $t\in \mathbb{R}_+$.

The Laplace transform $L_t^{\hat{x}}(s_1,..,s_j)$ of $Q_t^{\hat{x}}(\lambda_1,..,\lambda_j):= P^{\hat{x}}(t^{-1}\hat{x}_t[1_{A_\nu}]\leq\lambda_\nu;\nu=1,..,j)$ is given by

$$L_t^{\hat{x}} = \frac{F_t(\hat{x},\xi_t)-F_t(\hat{x},0)}{1 - F_t(\hat{x},0)} = 1 - \frac{1 - F_t(\hat{x},\xi_t)}{1 - F_t(\hat{x},0)},$$

$$\xi_t:=e^{-\xi/t}, \quad \xi:=\sum_{\nu=1}^{j} s_\nu 1_{A_\nu}.$$

Note that

$$t\Phi^*[1-\xi_t]\to\Phi^*[\xi], \quad t\to\infty.$$

Using this, it follows again from Lemma 7, Lemma 4, and (F.1) that

$$= \nu\delta(1-F_{\nu\delta}(\hat{x},\xi_{\nu\delta})) \to (1+\mu\Phi^*[\xi])^{-1}\Phi^*[\xi]\hat{x}[\varphi], \quad \mathbb{N}\ni\nu\to\infty.$$

From this by (5.2)

$$\lim_{\mathbb{N}\ni\nu\to\infty} L^{\hat{x}}_{\nu\delta}=(1+\mu\Phi^*[\xi])^{-1}, \ \delta>0.$$

The expression on the right is the Laplace transform of the limit d.f. proposed in (5.3). Denote this d.f. by Q_∞. By the continuity theorem $Q^{\hat{x}}_{\nu\delta}\to Q_\infty$, $\nu\to\infty$, and since Q_∞ is continuous, we have uniform convergence. Hence, we have convergence respective the metric

$$d(Q_1,Q_2):=\inf\{\varepsilon: Q_1(\lambda_1-\varepsilon,..,\lambda_j-\varepsilon)-\varepsilon\leq Q_2(\lambda_1,..,\lambda_j)$$
$$\leq Q_1(\lambda_1+\varepsilon,..,\lambda_j+\varepsilon)+\varepsilon, \lambda_\nu\in[0,\infty), \nu=1,..,j\},$$

defined for all pairs of j-dimensional distribution functions Q_1,Q_2 with $Q_1(0,..,0)=Q_2(0,..,0)=0$. Writing

$$Q^{\hat{x}}_t(\lambda_1,..,\lambda_j)=\sum_{\substack{n_\nu\leq t\lambda_\nu;\nu=1,..,j \\ n_1+..+n_j>0}} \frac{P^{\hat{x}}(\hat{x}_t[1_{A_\nu}]=n_\nu;\nu=1,..,j)}{P^{\hat{x}}(\hat{x}_t\neq\theta)},$$

it follows from Lemma 6 and (F.1) that $Q^{\hat{x}}_t$ is continuous in t>0 respective d. By the Croft-Kingman lemma [13] therefore

$$\lim_{\mathbb{R}_+\ni t\to\infty} d(Q^{\hat{x}}_t,Q_\infty)=0$$

which implies (5.3).

Concerning (5.4), note that

$$\hat{x}[M_t\xi]=E^{\hat{x}}\hat{x}_t[\xi]=P^{\hat{x}}(\hat{x}_t\neq\theta)E^{\hat{x}}(\hat{x}_t[\xi]\,|\,\hat{x}_t\neq\theta),$$

and apply (M) and (5.2).

6. THE SUPERCRITICAL CASE

By the martingale convergence theorem there exists a random variable W with $E^{\hat{x}}W\leq\hat{x}[\varphi]$ such that

$$W=\lim_{t\to\infty}\rho^{-t}\hat{x}_t[\varphi] \quad \text{a.s. } [P^{\hat{x}}].$$

Theorem 4 ([1]). Suppose $\rho>1$. Then $1-q\in D_0^+$, and for every almost everywhere continuous $\eta\in B$

$$\lim_{t\to\infty} \rho^{-t}\hat{x}_t[\eta]=\Phi^*[\eta]W \qquad \text{a.s.}[P^{\hat{x}}].$$

We have $E^{\hat{x}}W=\hat{x}[\varphi]$, $\hat{x}\in\hat{X}$, if and only if for some (and thus all) $t>0$

(X LOG X) $\Phi^*[E^{\langle\cdot\rangle}\hat{x}_t[\varphi]\log\hat{x}_t[\varphi]]<\infty$,

otherwise $W=0$ a.s. $[P^{\hat{x}}]$.

Remark. There exist a normalization sequence $\gamma_t=L(\rho^{-t})\rho^{-t}$, $L(s)$ slowly varying as $s\to 0$, and a random variable $\widetilde{W}$ such that

$$P^{\hat{x}}(\widetilde{W}<\infty)=1, \quad P^{\hat{x}}(\widetilde{W}=0)=\widetilde{q}(\hat{x}), \quad \hat{x}\in\hat{X},$$

and for every almost everywhere continuous $\eta\in B$

(6.1) $$\lim_{t\to\infty}\gamma_t\hat{x}_t[\eta]=\Phi^*[\eta]\widetilde{W} \qquad \text{a.s.}[P^{\hat{x}}].$$

To obtain (6.1) for $\eta=\varphi$, extend [9] by use of (M), (R), and Sevastyanov's transformation. To get from there to (6.1) with a general η, proceed as below, but with $\beta_t=\hat{x}_t[\varphi]$. A detailed treatment of this and other problems will be given in a separate paper.

Proof. This proof differs in parts from the proof given in [1]. For the moment fix $t>0$. By (FM), (M) with $\rho>1$, and (R) we can find an $\varepsilon>0$ such that $\Phi^*[1-F_t[1-\xi]]>\Phi^*[\xi]$ whenever $\|\xi\|<\varepsilon$. Suppose $\Phi^*[1-q]=0$. Then $\Phi^*[1-F_s[0]]\to 0$, as $s\to\infty$. By (F.2), (FM), (RM), and (M) there must then exist an $s>0$ such that $\|1-F_s[0]\|<\varepsilon$ and consequently $\Phi^*[1-F_{t+s}[0]]>\Phi^*[1-F_s[0]]$. But this contradicts the fact that $F_s[0]$ is non-decreasing. Hence, $q<1$ on a set of positive measure. From (IF) and $q=F_t[q]$, $t>0$,

$$1-q=T_t^0(1-q)+\int_0^t T_s^0\{k(1-f[q])\}ds.$$

By (2.1) and the irreducibility assumption on m, iteration of this equation yields q<1 on X, and using $T_s^0 B \subseteq C_0^0$, s>0, and (2.3-5) we get $1-q \in D_0^+$.

Next we turn to the degeneracy question for W. Define

$$\psi_t(\lambda)(x) := E^{\langle x\rangle} \exp\{-\rho^{-t}\hat{x}_t[\varphi]\lambda\} = F_t[\exp\{-\rho^{-t}\varphi\lambda\}](x),$$

$$\psi(\lambda)(x) := E^{\langle x\rangle} e^{-W\lambda}, \quad \lambda \geq 0,\ x \in X.$$

Then

(6.2) $$\psi_{t+s}(\lambda) = F_t[\psi_s(\rho^{-t}\lambda)], \quad t,s>0,\ \lambda \geq 0,$$

$$\psi(\lambda) = F_t[\psi(\rho^{-t}\lambda)], \quad \lambda>0,\ t>0.$$

The last equation implies

$$E^{\langle x\rangle} W = M_t[\rho^{-t} E^{\langle\cdot\rangle} W](x), \quad x \in X,\ t>0.$$

By (M) we therefore have either $E^{\langle x\rangle}W = \varphi(x)$, $x \in X$, or $E^{\langle x\rangle}W=0$, $x \in X$. Given this alternative, the first occurs if and only if

(6.3) $$\lim_{n\to\infty} \Phi^*[1-\psi_n(1)]>0.$$

We show that (6.3) is equivalent to (X LOG X). By (FM) and (6.2)

$$\Phi^*[1-\psi_n(1)] = \Phi^*[1-F_1[\psi_{n-1}(\rho^{-1})]]$$

$$= \rho\Phi^*[1-\psi_{n-1}(\rho^{-1})]\left\{1-\Phi^*\left[R_1(\psi_{n-1}(\rho^{-1}))\ \frac{1-\psi_{n-1}(\rho^{-1})}{\Phi^*[1-\psi_{n-1}(\rho^{-1})]}\right]\rho^{-1}\right\}$$

$$= \rho^{n-1}\Phi^*[1-\psi_1(\rho^{-n+1})]\prod_{\nu=1}^{n-1}\left\{1-\Phi^*\left[R_1(\psi_{n-\nu}(\rho^{-\nu}))\ \frac{1-\psi_{n-\nu}(\rho^{-\nu})}{\Phi^*[1-\psi_{n-\nu}(\rho^{-\nu})]}\right]\rho^{-1}\right\}.$$

Using (FM), (M), and (R),

$$\lim_{n\to\infty} \rho^{n-1}\Phi^*[1-\psi_1(\rho^{-n+1})]=1,$$

and there exist $\varepsilon>0$, $\varepsilon'>0$, and $n'>0$ such that

$$1-\varepsilon\rho^{-\nu}\varphi \leq \psi_{n-\nu}(\rho^{-\nu}) \leq 1-\varepsilon'\rho^{-\nu}\varphi, \quad n \geq n',\ \nu \leq n.$$

Hence, (6.3) is equivalent to

$$\sum_{\nu=1}^{\infty} \Phi^*[R_1(1-\vartheta\rho^{-\nu}\varphi)\varphi]<\infty$$

with some $\vartheta>0$. Now recall Lemma 2.

Lemma 8. For $0<\delta\in\mathbb{R}_+$ let $Y_{n,i}^{\delta}, Z_{n,i}^{\delta}, \beta_n^{\delta}$, $i=1,..,\hat{x}_{n\delta}[1]$, $n=0,1,2,..,$ be random variables such that

$$0\le Y_{n,i}^{\delta}\le Z_{n,i}^{\delta}, \quad \beta_n^{\delta}\ge 0 \qquad \text{a.e.}[P^{\hat{x}}].$$

Suppose the $Y_{n,i}^{\delta}$ are independent conditioned on $F_{n\delta}$, the same is true of the

$$\tilde{Y}_{n,i}^{\delta}:=Y_{n,i}^{\delta}1_{\{Z_{n,i}^{\delta}\le\beta_{n-1}^{\delta}\}}, \quad i=1,..,\hat{x}_{n\delta}[1],$$

and the distribution $G_{\langle x\rangle}^{\delta}$ of $Z_{n,i}^{\delta}$ depends only on $\langle x_i\rangle:=\hat{x}_{n\delta}^{n\delta,i}$,

$$\Phi^*[\int^{\infty}\lambda dG_{\langle\cdot\rangle}^{\delta}(\lambda)]<\infty.$$

Suppose further β_n^{δ} is $F_{n\delta}$-measurable, $\{\beta_n^{\delta}>0\}\supset\{\beta_{n+1}^{\delta}>0\}$,

(6.4) $\lim_{n\to\infty} (\beta_n^{\delta})^{-1}\beta_{n+1}^{\delta}>1$ a.e. on $\Gamma_{\delta}:=\bigcap_{n\in\mathbb{N}}\{\beta_n^{\delta}>0\}$,

and $(\beta_n^{\delta})^{-1}\hat{x}_{n\delta}[\varphi]1_{\{\beta_n^{\delta}>0\}}$ is bounded a.e. $[P^{\hat{x}}]$. Define

$$S_n^{\delta}:=1_{\Gamma_{\delta}}(\beta_{n-1}^{\delta})^{-1}\sum_{i=1}^{\hat{x}_{n\delta}[1]} Y_{n,i}^{\delta}, \quad \tilde{S}_n^{\delta}:=1_{\Gamma_{\delta}}(\beta_{n-1}^{\delta})^{-1}\sum_{i=1}^{\hat{x}_{n\delta}[1]} \tilde{Y}_{n,i}^{\delta}$$

Then

$$\lim_{n\to\infty}\{S_n^{\delta}-E^{\hat{x}}(\tilde{S}_n^{\delta}|F_{n\delta})\}=0 \qquad \text{a.s.}[P^{\hat{x}}].$$

Proof. Omitting the superscripts $\hat{x}$ and δ, setting $\delta=1$ elsewhere, and using (1.3), (M), and (6.4),

$$\sum_{n=1}^{\infty} E\{[\tilde{S}_n-E(\tilde{S}_n|F_n)]^2|F_{n-1}\}$$

$$\le \sum_{n=1}^{\infty} E\left\{(\beta_{n-1})^{-2}\sum_{i=1}^{\hat{x}_n[1]} E(\tilde{Y}_{n,i}^2|F_n)|F_{n-1}\right\}1_{\{\beta_{n-1}>0\}}$$

$$= \sum_{n=1}^{\infty}(\beta_{n-1})^{-2}\hat{x}_{n-1}[M_1[\int_0^{\beta_{n-1}}\lambda^2 dG_{\langle\cdot\rangle}(\lambda)]]1_{\{\beta_{n-1}>0\}}$$

$$\leq C_1 \sum_{n=1}^{\infty} \beta_{n-1}^{-1} \int_0^{\beta_{n-1}} \lambda^2 d\Phi^*[G\langle\cdot\rangle(\lambda)]1_{\{\beta_{n-1}>0\}}$$

$$\leq C_2 \int_0^{\infty} \lambda d\Phi^*[G\langle\cdot\rangle(\lambda)]+C_3,$$

$$\sum_{n=1}^{\infty} P\{S_n \neq \tilde{S}_n | F_{n-1}\}$$

$$= \sum_{n=1}^{\infty} E\{\sum_{i=1}^{\hat{x}_n[1]} P(Y_{n,i}>\beta_{n-1}|F_n)|F_{n-1}\}1_{\{\beta_{n-1}>0\}}$$

$$\leq \sum_{n=1}^{\infty} \hat{x}_{n-1}[M_1[\int_{\beta_{n-1}}^{\infty} dG\langle\cdot\rangle(\lambda)]]1_{\{\beta_{n-1}>0\}}$$

$$\leq C_4 \sum_{n=1}^{\infty} \beta_{n-1} \int_{\beta_{n-1}}^{\infty} d\Phi^*[G\langle\cdot\rangle(\lambda)]1_{\{\beta_{n-1}>0\}}$$

$$\leq C_5 \int_0^{\infty} \lambda d\Phi^*[G\langle\cdot\rangle(\lambda)]+C_6.$$

The $C_1,\ldots,C_6$ are finite, but in general random. Chebychev's inequality and the conditional Borel-Cantelli lemma complete the proof.□

Proof of Theorem 4 continued. For $\eta \in B_+$, $0<\delta \in \mathbb{R}_+$, and $n,r \in \mathbb{N}$ set

$$Y_{n,i}^{\delta} := Z_{n,i}^{\delta} := \hat{x}_{(n+r)\delta}^{n\delta,i}[\eta], \quad \beta_n^{\delta} := \rho^{(n+1)\delta}.$$

In the notation of Lemma 8,

$$\rho^{-(n+r)\delta}\hat{x}_{(n+r)\delta}[\eta] = \rho^{-r\delta}\{S_n^{\delta} - E^{\hat{x}}(\tilde{S}_n^{\delta}|F_{n\delta}) + \rho^{-r\delta}E^{\hat{x}}(S_n^{\delta}|F_{n\delta}) - \rho^{-r\delta}\varepsilon_n^{\delta},$$

$$\varepsilon_n^{\delta} := E^{\hat{x}}(S_n^{\delta} - \tilde{S}_n^{\delta}|F_{n\delta}).$$

By (1.3) and (M)

$$(1-\rho^{-r\delta}\alpha_{r\delta})\Phi^*[\eta]\rho^{-n\delta}\hat{x}_{n\delta}[\varphi] \leq \rho^{-r\delta}E^{\hat{x}}(S_n^{\delta}|F_{n\delta})$$

$$\leq (1+\rho^{-r\delta}\alpha_{r\delta})\Phi^*[\eta]\rho^{-n\delta}\hat{x}_{n\delta}[\varphi].$$

That is, if $\varepsilon_n^{\delta} \to 0$ a.s., $n \to \infty$, for every r, then by Lemma 8

$$\lim_{n\to\infty} \rho^{-n\delta}\hat{x}_{n\delta}[\eta]=\Phi^*[\eta]W \quad \text{a.s.}[P^{\hat{x}}].$$

We now prove $\varepsilon_n^\delta\to 0$. First, note that

$$\varepsilon_n^\delta \le \rho^{-n\delta}\hat{x}_{n\delta}[E^{\langle\cdot\rangle}\hat{x}_{r\delta}[\eta]]$$

$$\le(\rho^{r\delta}+\alpha_{r\delta})\Phi^*[\eta]\rho^{-n\delta}\hat{x}_{n\delta}[\varphi],$$

so that in any case

$$\lim_n \sup \rho^{-n\delta}\hat{x}_{n\delta}[1]<\infty \quad \text{a.s.}$$

Secondly

$$\varepsilon_n^\delta \le \|\eta\|\rho^{-n\delta}\hat{x}_{n\delta}[1]\sup_x \int_{\rho^{n\delta}} \lambda dP^{\langle x\rangle}(\hat{x}_{r\delta}[1]\le\lambda).$$

From (IF), for $y>1$,

$$\int_y^\infty \lambda dP^{\langle x\rangle}\ (\hat{x}_t[1]\le\lambda)=\int_0^t T_s^0\{kN_{t-s}^y\}(x)ds,$$

$$N_{t-s}^y(x):=\int_{\hat{X}}\sum_{n\ge y} n\pi(x,d\hat{x})P^{\hat{x}}(\hat{x}_{t-s}[1]=n).$$

We have $N_s^y(x)\le m[M_s[1]]\le\|m\|e^{\|km\|s}$ and $N_s^y(x)\to 0$, $y\to\infty$, for all x and s. Using $\|T_s^0\|\le 1$, $s>0$, boundedness of $p_s(x,y)$ on $[\varepsilon,t]\otimes X\otimes X$ for every $\varepsilon>0$, and dominated convergence, this implies

$$\sup_x \int_y^\infty \lambda dP^{\langle x\rangle}(\hat{x}_{r\delta}[1]\le\lambda)\to 0,\ y\to\infty.$$

Hence, $\varepsilon_n^\delta\to 0$, $n\to\infty$, for every r.

<u>Lemma 9</u>. If $\{\beta_t, t\in\mathbb{R}_+\}$ is a rightcontinuous process such that the $\beta_n^\delta:=\beta_{n\delta}$, $n\in\mathbb{N}$, $\delta>0$, satisfy the assumptions of the preceding lemma, with

$$\lim_{t\to\infty}\beta_t^{-1}\beta_{t+s}=\beta^s \quad \text{a.s. on } \Gamma=\bigcap_{t\ge 0}\{\beta_t>0\},\ s>0,$$

and $\widetilde{W}$ a random variable such that

$$\lim_{\mathbb{N}\ni n\to\infty}\beta_{n\delta}^{-1}\hat{x}_{n\delta}[\xi]=\Phi^*[\xi]\widetilde{W} \quad \text{a.s. on } \Gamma \tag{6.5}$$

for every $\delta>0$ and $\xi\in B_+$, then

$$\lim_{t\to\infty} \beta_t^{-1}\hat{x}_t[\eta]=\Phi^*[\eta]\widetilde{W} \quad \text{a.s. on } \Gamma$$

for any almost everywhere continuous $\eta\in B$.

<u>Proof</u>. For every $U\in A$ define

$$\xi_U^\delta(x):=P^{\langle x\rangle}(\hat{x}_t[1_U]=\hat{x}_t[1]\forall t\in[0,\delta]).$$

Clearly, $\xi_U^\delta(x)\uparrow 1_U(x)$, as $\delta\downarrow 0$, for every $x\in X$. Set

$$Y_{n,i}^\delta:=1_{\{\hat{x}_t^{n\delta,i}[1_U]=\hat{x}_t^{n\delta,i}[1]\forall t\in[n\delta,(n+1)\delta]\}}, \quad Z_{n,i}^\delta=1.$$

Then

$$\hat{x}_t[1_U]\geq S_n^\delta, \quad t\in[n\delta,(n+1)\delta],$$

and by Lemma 8 and (6.5)

$$\text{(6.6)} \quad \liminf_t \beta_t^{-1}\hat{x}_t[1_U]\geq\beta^{-\delta}\liminf_n S_n^\delta$$

$$=\beta^{-\delta}\liminf_n E^{\hat{x}}(\widetilde{S}_n^\delta|F_{n\delta})=\beta^{-\delta}\liminf_n E^{\hat{x}}(S_n^\delta|F_{n\delta})$$

$$=\beta^{-\delta}\liminf_n \beta_{n\delta}^{-1}\hat{x}_{n\delta}[\xi_U^\delta]=\beta^{-\delta}\Phi^*[\xi_U^\delta]\widetilde{W}\uparrow\Phi^*[1_U]\widetilde{W},\delta\downarrow 0 \text{ a.e.on}\Gamma.$$

Next, set

$$Y_{n,i}^\delta=Z_{n,i}^\delta=\hat{x}_{(n+1)\delta}^{n\delta,i}[1]+\#\{t:\hat{x}_{t-}^{n\delta,i}[1]>\hat{x}_t^{n\delta,i}[1],n\delta<t\leq(n+1)\delta\}.$$

Then

$$E^{\langle x\rangle}Y_{0,1}^\delta\leq e^{\alpha\delta},\alpha=||k||\cdot(||m||+1),$$

and again by Lemma 8 and (6.5)

$$\text{(6.7)} \quad \limsup_t \beta_t^{-1}\hat{x}_t[1]\leq\beta^{-\delta}\limsup_n S_n^\delta$$

$$=\beta^{-\delta}\limsup_n E^{\hat{x}}(\widetilde{S}_n^\delta|F_{n\delta})\leq\beta^{-\delta}\limsup_n E^{\hat{x}}(S_n^\delta|F_{n\delta})$$

$$\leq e^{\alpha\delta}\limsup_n \beta_{n\delta}^{-1}\hat{x}_{n\delta}[1]=e^{\alpha\delta}\Phi^*[1]\widetilde{W}\downarrow\Phi^*[1]\widetilde{W} \quad \text{a.e.}\Gamma,\ \delta\downarrow 0.$$

From (6.6) and (6.7) with $U=X$

$$\lim_{t\to\infty} \beta_t^{-1}\hat{x}_t[1]=\Phi^*[1]\widetilde{W} \quad \text{a.e. on } \Gamma,$$

and from this and (6.6) for any U with a boundary of measure zero

$$\limsup_t \beta_t^{-1}\hat{x}_t[1_U]=\Phi^*[1]\widetilde{W} - \liminf_t \beta_t^{-1}\hat{x}_t[1_U]$$

$$\leq\Phi^*[1_U]\widetilde{W} \quad \text{a.e. on } \Gamma.$$

Now take an appropriate denumerable class of such U's and apply Theorem 2.2 of [2]. □

REFERENCES

[1] Asmussen,S. and Hering,H. Strong limit theorems for general supercritical branching processes with applications to branching diffusions. Z.Wahrscheinlichkeitstheorie verw. Geb. 36, 195-212 (1976).

[2] Billingsley,P. Convergence of Probability Measures, John Wiley and Sons, New York, 1968.

[3] Harris,T.E. The Theory of Branching Processes, Springer-Verlag, Berlin-Göttingen-Heidelberg, 1963.

[4] Hering,H. Limit theorem for critical branching diffusion processes with absorbing barriers. Math.Biosci., 19, 355-37o (1974).

[5] Hering,H. Subcritical branching diffusions. Compositio Math., 34, 289-3o6 (1977).

[6] Hering,H. Refined positivity theorem for semigroups generated by perturbed differential operators of second order with an application to Markov branching processes. Math. Proc. Cambr. Phil. Soc. 83, 253-259 (1978).

[7] Hering,H. Uniform primitivity of semigroups generated by perturbed elliptic differential operators. Math. Proc. Cambr. Phil. Soc. 83, 261-268 (1978).

[8] Hering,H. Minimal moment conditions in the limit theory for general Markov branching processes. Ann. Inst. H. Poincaré, Sec. B, XIII, 299-319 (1977).

[9] Hoppe,F. Supercritical multitype branching processes. Ann. Probability, 4, 393-4ol (1976).

[10] Ikeda,N., Nagasawa,M., and Watanabe,S. Branching Markov processes I, II, III. J. Math. Kyoto Univ. 8, 233-278, 365-410 (1968); 9, 95-160 (1969).

[11] Itô,S. Fundamental solutions of parabolic differential equations and boundary value problems. Jap. J. Math., 27, 55-102 (1957).

[12] Joffe,A. and Spitzer,F. On multitype branching processes with $\rho \leq 1$. J. Math. Anal. Appl., 19, 409-430 (1967).

[13] Kingman,J.F.C. Continuous-time Markov processes. Proc. London Math. Soc., 13, 593-604 (1963).

[14] Krasnosel'skiǐ,M.A. Positive Solutions of Operator Equations, P.Noordhoff, Groningen, 1964 .

[15] Kreǐn,M.G. and Rutman,M.A. Linear operators leaving invariant a cone in a Banach space. Uspeki Matem. Nank. 3, 1-95 (1948); Amer. Math. Soc. Tranl. (1), 26 (1950).

[16] Sato,K. and Ueno,T. Multidimensional diffusion and the Markov process on the boundary. J. Math. Kyoto Univ., 4, 529-6o5 (1965).

[17] Savits,T. The explosion problem for branching Markov process. Osaka J. Math., 6, 375-395 (1969).

[18] Westcott,M. The probability generating functional. J. Austral. Math. Soc., 14, 448-466 (1972).

ANALYTICAL METHODS
FOR DISCRETE BRANCHING PROCESSES

F.M. HOPPE AND E. SENETA*

Cornell University, Ithaca, New York and Virginia Polytechnic Institute and State University, Blacksburg, Virginia

INTRODUCTION

A previous exposition [63] has expressed our view that simple branching processes may very successfully be studied in terms of real-variable transforms, as problems in functional iteration and

---

* Permanent addresses: F.M.Hoppe, University of Alberta, Edmonton. E.Seneta, Australian National University, Canberra.

in the structure of the functional equations which arise therefrom. Since that time, this view has been reinforced by two further factors in particular. The first is that there is an intimate and natural connection, induced by such functional equations themselves in the continuing absence of artificial prior assumptions, between the theory of simple branching processes and the theory of regularly varying functions [77]. The second is that the one-type theory extends, along the same lines, to the multitype case, to a degree of generality which subsumes the one-type case. Indeed, in the multitype case the scalar notion of regular variation (along a ray) still suffices, in a manner reminiscent of another (now well-known) feature of certain limit theorems in this setting, namely, directional convergence. (In both cases, such behaviour is ultimately induced by a Perron-Frobenius projection property of infinite products of perturbed non-negative matrices [73].)

The main purpose of this paper is to give, in its Part 1, a short unified development from this viewpoint of the theory of the multitype Bienaymé-Galton-Watson process. This results in an entity distinct from sections treating such processes in the monographs [1], [21], [32], [53], [83]. The emphasis is on new proofs, involving in particular the idea of linearization of a functional equation, whereby certain questions involving processes with arbitrary multitype offspring distributions are reduced to their counterparts with one-type linear offspring p.g.f.. Essentially the only new result, presented (in section 1.3) because it fits naturally into the framework, is the resolution of an old problem on the domains of attraction of diverse subcritical limit laws generated by initial distributions.

A secondary purpose is to introduce a further connection between real-variable and branching process theory. That is, to show that a certain generalization of regular variation, namely R-O variation [77], too, has a completely natural and dual, if rather less important, role to play in the theory of discrete branching processes. Since this material appears here for the first time in unified form (in Part 2) it is presented in a one-type setting, apart from several incidental allusions in Part 1.

The branching process monographs already mentioned, in conjunction with [82], give a rather extensive bibliographic survey of all branching process literature (at least to 1972), and we have not felt it necessary to aim at any measure of bibliographic completeness in ours. We thus cite, as in [63], mainly material which impinges on our sketch of the subject.

It should be mentioned at this stage that there are at least two kinds of important problems, even within discrete branching process theory, which apparently cannot be handled within the real variable framework to which we generally confine ourselves. The first is that of almost-sure convergence in limit theorems for supercritical processes, which is generally treated by martingale methods [23], [66], [75], [7]. The second is well-exhibited in the analytical work of S. Dubuc, particularly in his detailed investigations centering on the density of the limit variable in the supercritical case [11]-[14], although even here real-variable regular variation properties play an important role [12],[14]. (The apparent reason why one may not study such details of non-negative random variables from their real-variable Laplace transforms is the lack of suitable inversion formulae, no longer an obstacle if one resorts to characteristic functions/complex-valued Laplace transforms). While it is self-evident that treatment of such problems forms an important part of the analytical method for branching processes, we shall deal only with the first here.

PART 1

1.1 THE APPROACH

The early [22], [61]-[63] proofs for existence and uniqueness of limiting distributions and invariant measures utilized results of Kuczma [44]-[46] on monotonic and convex solutions of the Abel Schröder type functional equations. Such results do not, in general as yet, exist in the multidimensional setting; but, in any case, such analogues may be bypassed. Indeed it may be more natural to do so for a number of reasons, associated with the probabilistic structure which may be enumerated as follows:

(1) In one dimension, Kuczma's results are needed only for uniqueness, existence of solutions being verified directly.

(2) Uniqueness, when it obtains, generally follows once regular variation has been deduced. When it does not obtain, assumption of regular variation may ensure it.

(3) In more than one dimension, even consideration of existence of an appropriate solution may be thought as hinging on regular variation.

These observations explain our focus on regular variation as a dominant feature in questions of existence and uniqueness.

Our notation will conform with [24] and [26] and we state the bare essentials. $F(x)$ is the offspring p.g.f. of a positively regular (and non-singular in case of criticality) Galton-Watson process $\{Z_n = (Z_n^{(1)},\ldots,Z_n^{(d)})\}$ taking values in $\mathbf{Z}^d$ and whose offspring expectation matrix M has a maximal eigenvalue $0 < \rho < \infty$ with corresponding left and right eigenvectors v and u respectively, uniquely specified by the scalar product normalizations $v\cdot u = 1\cdot u = 1$. $F_n(x)$ is the nth functional iterate of F and we shall denote $F_n(0)$ by F_n. x and s will always refer to generic vectors and scalars respectively. Inequalities and functional operations between vectors are taken componentwise and if $x \in \mathbf{R}^d$ and $k \in \mathbf{Z}^d$ then $\mathbf{x}^k = \Pi_{i=1}^d x_i^{k_i}$. We shall denote by e_i $(1 \le i \le d)$ the point in $\mathbf{Z}^d$ with all coordinates zero except for a one in position i. $Z \log Z$ will mean that $E[Z_1^{(j)} \log Z_1^{(j)} | Z_0 = e_i]$ is finite, for all $1 \le i,j \le d$.

It will be necessary in the sequel to apply some convergence properties of the iterates $F_n(x)$. These follow from the expansion

$$1 - F(x) = (M-E(x))(1-x) \tag{0}$$

established by Joffe and Spitzer [33] and are for $\rho \leq 1$ with $x \neq 1$:

$$\lim_{n\to\infty} (1-F_n(x))/v\cdot(1-F_n(x)) = u; \tag{1a}$$

and

$$\lim_{n\to\infty} v\cdot(1-F_{n+1}(x))/v\cdot(1-F_n(x)) = \rho. \tag{1b}$$

When $\rho = 1$ it will be useful to know that [29]:

$$\lim_{n\to\infty} (F_n(x)-F_n(y))/v\cdot(F_n(x)-F_n(y)) = u \tag{2a}$$

for $x \neq y \neq 1$; and for $x \neq 1$

$$\lim_{n\to\infty} v\cdot(F_{n+1}(x)-F_n(x))/v\cdot(F_{n+1}-F_n) = 1. \tag{2b}$$

In connection with discrete limit laws and invariant measures, we restrict ourselves to the non-supercritical case since , if $\rho > 1$ and the extinction probability vector q has all the components strictly positive, analogous results hold as a consequence of the transformation $F(x) \to F(xq)/q$ taking a supercritical p.g.f. into a subcritical one. We leave it to the interested reader to make the appropriate transformations.

The order of presentation and the functional equations to be considered are as follows:

(a) $1-A(F(x)) = \rho(1-A(x))$, characterizing the conditional Yaglom limit distribution;

(b) $1-B(F(x)) = \rho^{\alpha}(1-B(x))$, occurring in connection with Yaglom limits arising as a result of varying the initial distributions;

(c) $G(F(x)) = G(x) + 1$, determining invariant measures for the process without immigration;

(d) $P(x) = B(x)\,P(F(x))$, determining invariant measures for the process with immigration;

(e) $\phi(\rho s) = F(\phi(s))$, characterizing the supercritical normed limit distribution.

1.2 THE SUBCRITICAL CASE AND THE SCHRÖDER EQUATION

Theorem A.

If $\rho < 1$ then

$$\lim_{n\to\infty} \frac{v\cdot(F_n(x)-F_n)}{v\cdot(1-F_n)} \tag{3}$$

exists and is the unique p.g.f. solution to the Schröder equation

$$1 - A(F(x)) = \rho(1-A(x)), \quad A(0) = 0. \tag{4}$$

Moreover,

$$1 - A(1-su) = sL(s) \tag{5}$$

where L is slowly varying at 0.

Proof.

Letting $A_n(x) = v\cdot(F_n(x)-F_n)/v\cdot(1-F_n)$ it results that $1-A_n(F(x)) = B_n(1-A_n(x))$ where $B_n = v\cdot(1-F_{n+1}(x))/v\cdot(1-F_n(x))$. If next we write $A(x) = \limsup\ (n\to\infty)\ A_n(x)$, since $B_n \to \rho$ equation (4) obtains for A.

If $Q(x)$ denotes any convex and non-decreasing solution (such as the one just constructed for instance) we can define for $0 \le s \le 1$ and n a positive integer, the sequence of functions $\{h_n(s)\}$ by $h_n(s) = \rho^{-n}[1-Q(1-s(1-F_n))]$. Because F is convex, $F(1-s(1-F_n)) \le 1-s(1-F_{n+1})$ and since Q is non decreasing $h_{n+1}(s) \le h_n(s)$, so that the limit $h(s)$ exists, is concave non-decreasing on $[0,1]$ with the obvious boundary values $h(0) = 0$, $h(1) = 1$.

Equations (1a) and (1b) imply that for arbitrarily small positive ε and n sufficiently large, $(1-\varepsilon)\rho(1-F_{n+1}) \le 1-F_n \le (1+\varepsilon)\rho(1-F_{n+1})$ giving $h_n((1-\varepsilon)s\rho) \le h_n(s) \le h_n((1+\varepsilon)s\rho)$. Passage to the limit and continuity produces the equation $h(\rho s) = \rho h(s)$. Since h is concave with $h(0) = 0$ it is evident that $h(s)/s$ is constant and the constant must be 1 to satisfy $h(1) = 1$.

Summarizing, we have demonstrated that

$$\lim_{n\to\infty} \rho^{-n}[1-Q(1-s(1-F_n))] = s, \tag{6}$$

and we shall use this relation to force uniqueness. Fix x such that $A(x) \neq 0$, and let $\{n'\}$ denote a subsequence of the integers for which $\limsup A_n(x)$ is achieved. Again, if ε is small and n sufficiently large then from the definition of $A(x)$ and (1a), $(1-\varepsilon)(1-A(x))(1-F_{n'}) \le 1-F_{n'}(x) \le (1+\varepsilon)(1-A(x))(1-F_{n'})$ and arguing as above we get $h_{n'}((1-\varepsilon)(1-A(x))) \le \rho^{-n'}(1-Q(F_{n'}(x)) \equiv 1-Q(x) \le h_{n'}((1+\varepsilon)(1-A(x)))$, resulting by (6) in $1-A(x) = 1-Q(x)$, that is uniqueness.(The case $A(x) = 0$ is handled analogously.)

Uniqueness now shows that from any subsequence of $\{A_n(x)\}$ we can extract a sub-subsequence converging to the same limit A. The entire sequence thus converges, settling the first part of the theorem.

Focussing on (5), letting $a_n = v\cdot(1-F_n)$ and introducing $\phi(s) = 1-A(1-su)$, the bounds $s(1-\varepsilon)(1-F_n) \le sa_n u \le s(1+\varepsilon)(1-F_n)$ in conjunction with (6) show that $\rho^{-n}\phi(a_n s) \to s$ as $n \to \infty$, and so $\phi(a_n s)/\phi(a_n) \to s$ as $n \to \infty$. This is equivalent to (5) in view of the following lemma of Rubin and Vere-Jones [59]: Suppose that a function ϕ is monotonic increasing in $[0,a)$ and that $\{\theta_n\}$ is a sequence of positive reals tending to 0 in such a way that $\theta_n/\theta_{n+1} \le c < \infty$. Then $\lim(n \to \infty)\phi(\theta_n s)/\phi(\theta_n) = s^\alpha$ iff ϕ is regularly varying of index α. □

Corollary (Yaglom limit).

If $Z_0 = i$, then

$$\lim_{n\to\infty} P[Z_n=k \mid Z_n\neq 0] = a(k)$$

where $\{a(k)\}$ is a probability distribution independent of i with p.g.f. $A(x)$.

Proof.

The conditional p.q.f. at time n is just $1-(1-F_n(x)^i)/(1-F_n^i)$ which tends to $A(x)$ since from (1a) (see lemma 2 of [33])

$(1-F_n(x)^i)/v\cdot(1-F_n(x)) \to i\cdot u$ independent of x. □

Versions of the Yaglom limit theorem under various moment assumptions date back to 1947. The multitype case under the above conditions may be found in [33] using a different method (the results in Theorem A are not given there). The uniqueness characterization of $\{a(k)\}$ first appeared in [22] and (5) is contained in [70], [24].

Equation (5) gives some information on the tail of $\{a(k)\}$. Let W be a random vector with p.g.f. $A(x)$, and let $\psi(s) = E[\exp(-sW u)] \equiv A(e^{-su})$ denote the L.S. transform of $W\cdot u$. By writing $e^{-s} = 1-s+o(s)$ it obtains from (5) that $1-\psi(s) = sL_2(s)$ where $L_2(s) \sim L(s)$ as $s \to 0$. A standard Tauberian argument [16,page 445], [77] applies yielding

$$\int_0^s \Pr[W\cdot u > t]dt \sim L_2(s^{-1}),\ s \to \infty$$

where $L_2(s) \uparrow E[W\cdot u] \le \infty$ as $s \downarrow 0$. A density version of this theorem results in

$$P[W\cdot u > s] = o(s^{-1}L_2(s^{-1}))$$

and an estimate using the growth of slowly varying functions shows

$$E[(W\cdot u)^\tau] < \infty \qquad \text{for} \quad 0 \le \tau < 1.$$

In [33] it was proven that $E[W\cdot u] < \infty$ iff $Z \log Z$ holds in which circumstance there is a positive constant γ such that $\rho^{-n}v\cdot(1-F_n) \to \gamma$ as $n \to \infty$. The limit always exists but $\gamma \equiv 0$ iff this condition is violated and we now investigate the behaviour of the extinction probability in general.

First of all by letting $s \to 0$ in (5) through the sequence $\{a_n\}$ we get $1-A(1-a_nu) = a_nL(a_n)$. From (1a) there is a positive null sequence $\{\varepsilon_n\}$ such that $(1-\varepsilon_n)a_nu \le 1-F_n \le (1+\varepsilon_n)a_nu$ and by the uniform convergence property of slowly varying functions [43], [77] it follows that $1-A(1-a_nu) \sim 1-A(F_n) = \rho^n$. We therefore obtain

$$a_n \sim \rho^n L^{-1}(a_n) \quad \text{as} \quad n \to \infty.$$

This is not quite good enough because a_n appears as the argument of the slowly varying function on the right. To do better, let $\phi(s) = 1 - A(1-su)$, $0 \le s \le 1$. The inverse function ϕ^{-1} is convex and thus $\phi^{-1}(t)/t$ increases with t. Since $\phi(\rho s)/\phi(s) \to \rho$ (by (5)) as $s \to 0$, if s is chosen sufficiently small then $\phi(\rho s) \le (\rho+\varepsilon)\phi(s)$ where ε is such that $\rho+\varepsilon < 1$. Let $\lambda \in [\rho+\varepsilon, 1]$ obtaining $\phi(\rho s) \le (\rho+\varepsilon)\phi(s) \le \lambda\phi(s) \le \phi(s)$. Putting $t = \phi(s)$ we have

$$\frac{\phi^{-1}(\phi(\rho s))/\phi(\rho s)}{\phi^{-1}(\phi(s))/\phi(s)} \le \frac{\phi^{-1}(\lambda t)/\lambda t}{\phi^{-1}(t)/t} \le 1.$$

But the left side is just $\rho\phi(s)/\phi(\rho s)$ and approaches 1 as t (and therefore s) approaches 0, demonstrating that $\phi^{-1}(t) = tL_1(t)$ with L_1 slowly varying at 0. With the selection $t = \rho^n$ we conclude that $\phi^{-1}(\rho^n) = \rho^n L_1(\rho^n)$. But we have just shown above that $\phi(a_n) \sim \rho^n$ so by the uniform convergence property we are permitted to replace ρ^n by $\phi(a_n)$ in the left hand side of the previous equation and retain asymptotic equality. Doing this we are left with

$$a_n \equiv v\cdot(1-F_n) \sim \rho^n L_1(\rho^n) \quad \text{as} \quad n \to \infty.$$

It is clear that $L_1(t) \downarrow E[W\cdot u]^{-1}$ which is strictly positive iff $Z \log Z$ holds.

By comparing the expressions $a_n \sim \rho^n L^{-1}(a_n)$ and $a_n \sim \rho^n L_1(\rho^n)$, we see that L and L_1 are related through $L_1(\rho^n)\cdot L(\rho^n L_1(\rho^n)) \sim 1$. Actually more generally $L_1(s)L(sL_1(s)) \sim 1$, $s \to 0$, L and L_1 being conjugate in the sense of de Bruijn [6], [77].

For the one-dimensional versions of the above results, we refer the reader to [74]. Dubuc [11] has some related material.

1.3 EFFECT OF AN INITIAL DISTRIBUTION ON THE YAGLOM LIMIT

Clothed within the fabric of Theorem A is the following skeleton idea. Given a solution Q of the Schröder equation we are able to construct a function H of a scalar variable (put $H(s) = 1-h(1-s)$) by the formula

$$H(s) = \lim_{n\to\infty} 1-\rho^{-n}[1-Q(1-s(1-F_n))]$$

which satisfies a one-dimensional linearized equation

$$1-H(1-\rho+\rho s) = \rho(1-H(s)), \quad H(0) = 0$$

and such that

$$Q(x) = H(A(x)).$$

This procedure effectively linearizes the Schröder equation reducing it to a much simpler form. Of course, because there is uniqueness, the representation displayed by the last equation seems somewhat superfluous. In this section and, more significantly, in the next on invariant measures, this technique will be developed further to give new results and insight.

The set up we have in mind is the following. We consider a Galton-Watson process with an arbitrary initial distribution whose p.g.f. is π and look for a conditional Yaglom limit. Seneta and Vere-Jones [79] using $\pi(s) = 1-(1-s)^{\alpha}$, $0 < \alpha \le 1$, showed a proper limit to result with p.g.f. $1-(1-A(s))^{\alpha}$. Rubin and Vere-Jones [59] deduced that this limit obtained iff $1-\pi(1-s) = s^{\alpha}L(s)$ where L varies slowly at 0. The following was proven in [70] and we state the multitype version without proof.

<u>Theorem</u>.

If $\lim(n \to \infty)$ $P[Z_n = k | Z_n \ne 0, Z_0 \sim \pi]$ exists for all $k \ne 0$ (and is not entirely concentrated at infinity) then this limit distribution is proper and for some $0 < \alpha \le 1$ its p.g.f. must satisfy

$$1-B(F(x)) = \rho^{\alpha}(1-B(x)), \quad B(0) = 0. \tag{7}$$

Among all solutions of (7) (if more than one exists) there is a unique such satisfying a regular variation condition

$$1-B(1-su) = s^{\gamma}L(s), \qquad s \to 0$$

in which case $\gamma = \alpha$ and the solution is $1-(1-A(x))^{\alpha}$.

Equation (7) determines the totality of all conditional Yaglom limits that may arise when the initial distribution is permitted to vary, since if we take $\pi(x) = B(x)$ for a p.g.f. solution B, then the Yaglom limit exists and has p.g.f. B(x). When $\alpha = 1$ there is a unique solution. It was suggested in a footnote to [59] that for $0 < \alpha < 1$ uniqueness is also obtained. Theorem B shows that this is not the case. It is then desirable to characterize the class of initial p.g.f.'s which give rise to each possible limit. This is the content of Theorem C which therefore contains and generalizes the previously cited work on this topic.

Theorem B.

There is a 1-1 correspondence between p.g.f. solutions of (7) and invariant measures for the process with the same F.

Theorem C.

Suppose B(x) is a p.g.f. solution of (7) for some $0 < \alpha \leq 1$ (and thus a possible Yaglom limit p.g.f.). Then an initial distribution with p.g.f. $\pi(x)$ gives rise to the Yaglom limit distribution with p.g.f. B(x) iff

$$\frac{1 - \pi(1-su)}{1 - B(1-su)} = L(s) \tag{8}$$

where L is slowly varying at 0.

The proof of Theorem B proceeds by way of the following lemma asserting that it is sufficient to consider only a single type linear offspring p.g.f. process.

Lemma (linearization).

Let H(s) denote any p.g.f. solution of

$$1 - H(1-\rho+\rho s) = \rho^{\alpha}(1-H(s)), \qquad H(0) = 0. \tag{9}$$

Then there is a 1-1 correspondence between solutions of (7) and (9) given by

$$B(x) = H(A(x)). \tag{10}$$

Proof of lemma.

Substitution shows that a function defined by (10) satisfies (7). For the converse we must construct a p.g.f. solution of (9) giving the representation (10). The details are akin to the proof of (6) in Theorem A.

Define $h_n(s) = \rho^{-n\alpha}[1-B(1-s(1-F_n))]$. The limit $h(s)$ exists and satisfies the equation $h(\rho s) = \rho^{\alpha}h(s)$. Our required function $H(s) = 1-h(1-s)$ is evidently a p.g.f. being the limit of a sequence of p.g.f. and satisfies (9). Working with the inequalities $(1-\varepsilon)(1-A(x))(1-F_n) \leq 1-F_n(x) \leq (1+\varepsilon)(1-A(x))(1-F_n)$ we obtain $h_n((1-\varepsilon)(1-A(x)) \leq \rho^{-n\alpha}(1-B(F_n(x))) \equiv 1-B(x) \leq h_n((1+\varepsilon)(1-A(x)))$ giving in the limit $h(1-A(x)) = 1-B(x)$, affirming (10). □

Proof of Theorem B.

This lemma in hand, we may now pursue non uniqueness for $\alpha < 1$. First we adjoin the fact that in the subcritical case invariant measures are not unique. Although this topic will be covered in its own right in the next section, the result we need does not rest on anything contained hereafter pertaining only to the single type process with a linear offspring p.g.f. The g.f. of an invariant measure for such a process satisfies the Abel equation

$$D(1-\rho+\rho s) = D(s) + 1, \qquad D(0) = 0.$$

It is known [87] that g.f. solutions are not unique. Given such a g.f. D define $r(s) = \exp[(\alpha-1)D(s)\log\rho]$ and integrate to produce $R(s) = \int_0^s r(t)\,dt$. By Abel's lemma, R can be normalized to a proper p.g.f. if $R(1-) < \infty$. But it is not hard to check that $D(s)-\log(1-s)/\log\rho \leq 1$ implying $r(s) \leq \rho^{\alpha-1}(1-s)^{\alpha-1}$ which in turn gives $R(s) \leq \rho^{\alpha-1}\alpha^{-1}(1-(1-s)^{\alpha})$ and in particular $R(1-)<\infty$. It is now a straightforward verification that the function $H(s) = R(s)/R(1-)$ satisfies (8).

From each invariant measure we have just constructed a Yaglom limit as $B(x) = R(A(x))/R(1-)$. Conversely given a limit p.g.f. $H(s)$ then $D(s) \equiv \log(1-H(s))/\log\rho$ is checked to be the g.f. of an invariant measure. By Bernstein's theorem, we have the 1-1 correspondence. Actually we have only proved this correspondence in the case of a linear offspring p.g.f. (although non-uniqueness has been established through (10)), but it is generally true in view of the corresponding linearization lemma to be used in Theorem D showing there is a representation analogous to (10) for invariant measures. The reader is alerted to observe this at the proper place in the text. □

Remark.

Although there is a unique regularly varying solution of (7) all solutions are R-O varying (see Part 2 of this paper) in the sense that there exist constants $0 < c_1 \leq 1$ and $1 \leq c_2 < \infty$ such that if $\psi(s) = s^{-\alpha}(1-B(1-su))$ then for all $\lambda > 0$

$$c_1 \leq \liminf_{s\to 0} \psi(\lambda s)/\psi(s) \leq \limsup_{s\to 0} \psi(\lambda s)/\psi(s) \leq c_2.$$

In view of (5) and (10) it suffices to prove these inequalities for the function H. Observe that $(\rho s)^{-\alpha}h(\rho s) = s^{-\alpha}h(s)$ and bounding $s^{-\alpha}h(s)$ for $s \in [\rho,1]$ we have $\rho^{\alpha} \leq s^{-\alpha}h(s) \leq \rho^{-\alpha}$ for all $s \in (0,1]$ and the above inequalities hold with $c_1 = \rho^{2\alpha}$ and $c_2 = \rho^{-2\alpha}$ both independent of H.

Proof of Theorem C.

The p.g.f. of the conditional distribution $\{\Pr[Z_n=k|Z_n\neq 0, Z_0 \sim \pi]\}$ is $\pi_n(x) \equiv (\pi(F_n(x)) - \pi(F_n))/(1-\pi(F_n))$, so that $1-\pi_n(x) = (1-\pi(F_n(x)))/(1-\pi(F_n))$. Given $\varepsilon > 0$ for n sufficiently large, setting $a_n = v\cdot(1-F_n)$ and using (1a) and (3)

$$(1-\varepsilon)(1-A(x))a_n u \le 1-F_n(x) \le (1+\varepsilon)(1-A(x))a_n u$$

with analogous inequalities, 0 replacing x. We then deduce

$$1-\pi(1-(1-\varepsilon)(1-A(x))a_n u) \le 1-\pi(F_n(x)) \le 1-\pi(1-(1+\varepsilon)(1-A(x))a_n u).$$

Three similar inequalities then show that

$$\frac{1-\pi(1-(1-\varepsilon)(1-A(x))a_n u)}{1-B(1-(1+\varepsilon)(1-A(x))a_n u)} \frac{1-B(1-(1-\varepsilon)a_n u)}{1-\pi(1-(1+\varepsilon)a_n u)} \le \frac{1-\pi_n(x)}{1-B(x)}$$

$$\le \frac{1-\pi(1-(1+\varepsilon)(1-A(x))a_n u)}{1-B(1-(1-\varepsilon)(1-A(x))a_n u)} \frac{1-B(1-(1+\varepsilon)a_n u)}{1-\pi(1-(1-\varepsilon)a_n u)} .$$

If (8) holds then as $n \to \infty$ the left side goes to $(1-\varepsilon)^2/(1+\varepsilon)^2$ and the right side to $(1+\varepsilon)^2/(1-\varepsilon)^2$ and letting $\varepsilon \downarrow 0$ it follows that $1-\pi_n(x) \to 1-B(x)$ proving sufficiency of (8).

To argue the converse define for $s \in [0,1)$, $\phi(s) = (1-\pi(1-su))/((1-B(1-su))/s)$, which, being the ratio of an increasing function divided by a decreasing function, is increasing. (The value at 0 is defined by right-continuity but it is immaterial for the purpose at hand.) For x of the form $x = tu$, t a scalar in (0,1), if ε is an arbitrary small positive vector and n is sufficiently large, a familiar argument gives

$$1-F_n(x+\varepsilon) \le a_n(1-A(x))u \le 1-F_n(x-\varepsilon)$$

and

$$1-F_n(u+\varepsilon) \le a_n(1-A(u))u \le 1-F_n(u-\varepsilon).$$

This makes a sandwich

$$\frac{1-\pi_n(x+\varepsilon)}{1-\pi_n(u-\varepsilon)} \le \frac{1-\pi(1-a_n(1-A(x))u)}{1-\pi(1-a_n(1-A(u))u)} \le \frac{1-\pi_n(x-\varepsilon)}{1-\pi_n(u+\varepsilon)}$$

which compresses to

$$\lim_{n\to\infty} \frac{1-\pi(1-a_n(1-A(x))u)}{1-\pi(1-a_n(1-A(u))u)} = \frac{1-B(x)}{1-B(u)} .$$

Likewise

$$\lim_{n\to\infty} \frac{1-B(1-a_n(1-A(x))u)}{1-B(1-a_n(1-A(u))u)} = \frac{1-B(x)}{1-B(u)} .$$

By putting $\theta_n = a_n(1-A(x))$ and $s = (1-A(x))/(1-A(u))$ we see that $\lim (n\to\infty)\ \phi(\theta_n s)/\phi(\theta_n) = s$. The lemma of Rubin and Vere-Jones implies that $\phi(s) = sL(s)$ which is exactly (8). □

1.4 INVARIANT MEASURES AND THE ABEL EQUATION

An invariant measure for the Galton-Watson process without immigration is equivalent (up to constant multiples) to a g.f. solution of Abel's equation (see [25])

$$G(F(x)) = G(x) + 1, \qquad G(0) = 0. \tag{11}$$

(i) The Subcritical Case

Theorem D.

(a) For every probability measure ϑ on $[0,1)$ the function

$$G(x) = \int_0^1 \{ \sum_{n=-\infty}^{\infty} [\exp(-(1-A(x))\rho^{n-t}) - \exp(-\rho^{n-t})] \} \vartheta(dt)$$

is the g.f. of an invariant measure and conversely every invariant measure has such a representation.

(b) $$\lim_{s\to 0} \frac{(\log\rho)\ G(1-su)}{\log s} = 1$$

and if $\{\pi(k)\}$ is an invariant measure then

$$\sum_{k:k\cdot u\le s} \pi(k) \sim -\log s/\log\rho \quad \text{as} \quad s \to \infty.$$

The proof of this theorem is based on the following lemma.

Lemma.

Let $D(s)$ denote any g.f. solution of

$$D(1-\rho+\rho s) = D(s) + 1, \qquad D(0) = 0. \tag{12}$$

There is a 1-1 correspondence between solutions of (11) and (12) induced by

$$G(x) = D(A(x)). \tag{13}$$

Proof.

$D(A(x))$ obviously satisfies (11). For the converse we proceed as in the linearization lemma of the previous section. For any solution define functions $d_n(s)$ by $d_n(s) = G(1-s(1-F_n))-n$, $0 \le s \le 1$. The sequence $\{d_n(s)\}$ is non-decreasing in n and bounded above since for each s we may find an x such that $1-F_n(x) \le s(1-F_n)$ if n is large and thus $d_n(s) \le G(x)$. The limit $d(s)$ exists and can be shown to satisfy $d(\rho s) = d(s) + 1$. $D(s) = d(1-s)$ is then the desired function. (13) follows using the same bounds as for (10). □

Proof of Theorem D.

(a) It was shown in [87] that $D(s)$ has the representation

$$D(s) = \int_0^1 \{ \sum_{n=-\infty}^{\infty} [\exp(-(1-s)\rho^{n-t})-\exp(-\rho^{n-t})]\}\vartheta(dt).$$

Now invoke (14).

(b) For each s let n be such that $1-\rho^n \le s < 1-\rho^{n+1}$ resulting in

$$\frac{n}{n+1} \le \frac{\log D(s)}{\log(1-s)} \le \frac{n+1}{n}$$

so as $s \to 1-$, and consequently $n \to \infty$, the inner quantity approaches 1. If $\phi(s) = 1-A(1-su)$ then

$$\frac{(\log\rho)G(1-su)}{\log s} = \frac{(\log\rho)D(1-\phi(s))}{\log \phi(s)} \frac{\log \phi(s)}{\log s}$$

and the right hand side tends to 1 as $s \to 0$ since $\phi(s) = sL(s)$ implies $\log \phi(s)/\log s \to 1$. A Tauberian argument [16],[77] concludes the proof. □

Remarks.

For the single type process with $E[Z_1 \log Z_1] < \infty$ part (a) of this theorem was stated in [1] and attributed to an unpublished manuscript of Kesten and Spitzer mentioned by the latter at the conference where [87] was presented. There are indications of a proof in the textbook [1] but most of the details are banished to the problem set, in which connection see [86]. Part (b) was established by Lipow [52] again for the single type process with $E[Z_1 \log Z_1] < \infty$.

If ϑ is taken as Lebesgue measure, the corresponding g.f. is $G(x) = \log(1-A(x))/\log \rho$. It has the property that

$$\lim_{s\to 0} G(1-\lambda su) - G(1-su) = \log \lambda/\log \rho, \ \lambda > 0.$$

This relationship will be further remarked upon as the corollary to Theorem F.

(ii) The Critical Case

It was shown in [61] that there is a unique invariant measure arising from a sequence of normalized transition probabilities. The following multitype extension was established in [29]. The existence proof we give is new and based entirely on properties of the Abel equation.

Theorem E.

There is a unique g.f. solution to (11). It is given as

$$\lim_{n\to\infty} \frac{v\cdot(F_n(x) - F_n)}{v\cdot(F_{n+1} - F_n)} .$$

Remarks.

Let $G_n(x) = v\cdot(F_n(x)-F_n)/v\cdot(F_{n+1}-F_n)$. Then $G_n(F(x)) = G_n(x) + C_n$ where $C_n = v\cdot(F_{n+1}(x)-F_n(x))\ /v\cdot(F_{n+1}-F_n)$. If we write $G(x) = \limsup(n\to\infty)\ G_n(x)$ then since by (2b) $C_n \to 1$, the function G defines a solution to (11) which however is not necessarily a g.f. Let $\{n'\}$ denote any subsequence of the integers. We may also define $\hat{G}(x) = \limsup G_{n'}(x)$ and it is also a solution to

(11). That $\hat{G}(x) = G(x)$ and so, in fact, the limit exists, will follow on the basis of the next lemma.

Lemma.

Let $Q(x)$ be any convex and non-decreasing solution. Then for any $s \geq 0$

$$\lim_{n\to\infty} Q(F_n+s(F_{n+1}-F_n)) - n = s. \tag{14}$$

Note that for arbitrary s it is not true that $F_n + s(F_{n+1}$ $F_n)$ always lies in the domain of definition of Q. But from (2b) $v\cdot(F_{n+k}-F_n)/v\cdot(F_{n+1}-F_n) \to k$ as $n \to \infty$ so if $s \leq k-1$ then $F_n+s(F_{n+1}-F_n) \leq F_n+(F_{n+k}-F_n) = F_{n+k} < 1$ if n is sufficiently large. This qualification on the meaning (14) should be kept in mind in the ensuing discussion.

Proof of Lemma.

Let $H_n(s) = Q(F_n+s(F_{n+1}-F_n))-n$. For $0 \leq s \leq 1$ the argument of Q is a convex combination of F_n and F_{n+1}, giving on the basis of (11) that $Q(F_n+s(F_{n+1}-F_n))+1 = Q(F(F_n+s(F_{n+1}-F_n))) \leq Q(F_{n+1}+s(F_{n+2}-F_{n+1}))$ so $H_n(s) \leq H_{n+1}(s)$. Since for all s and n, $H_n(s) \leq H_n(1) = 1$ the limit $H(s)$ exists in $[0,1]$.

If $s \in (1,2)$ then $H_n(s)$ may also be written as $Q(F_{n+1} + (s-1)(F_{n+1}-F_n))-n$ and by (2a) and (2b) if ε is so small that $(1+\varepsilon)(s-1) < 1$, for all large n we have $H_{n+1}((1-\varepsilon)(s-1))+1 \leq H_n(s) \leq H_{n+1}((1+\varepsilon)(s-1))+1$. If we pass to the limit and use the continuity of H in $(0,1)$, valid because H is convex in $[0,1]$, we have the existence of the limit $H(s)$ for $s \in (1,2)$. If $s=2$ the same argument applies but using the (just established) continuity of H in $(0,2)$. By induction the limit $H(s)$ exists for all s and is a convex function.

The above analysis has also given us the functional equation $H(s+1) = H(s) + 1$. This shows that $H(k) = k$ whenever k is an integer and then convexity establishes that $H(s) = s$, proving (14).

□

Proof of Theorem E.

If $G(x)$ is the solution obtained by taking lim sup $G_n(x)$ and Q is an arbitrary (but convex monotonic solution) then by (2a) if ε is small and n' large $(1-\varepsilon)G(x)(F_{n'+1}-F_{n'}) \leq F_{n'}(x) - F_{n'} \leq (G(x)+\varepsilon)(F_{n'+1}-F_{n'})$ where $\{n'\}$ is a subsequence such that $G_{n'}(x) \to G(x)$ for this particular x. Using (14) and then letting $\varepsilon \downarrow 0$ we obtain $Q(x)=G(x)$. By the remarks after the statement of this theorem we see that the proof is complete. □

Remarks.

Although regular variation does not enter the critical picture, in so far as the present discussion is concerned, (It does appear in some unrelated work of Slack [84],[85]; see also [18],[19],[89].) the limit relation (14) may be thought of as a natural counterpart since it is formally similar to (6) which is equivalent to regular variation in the subcritical case.

Corollary [61].

Let T denote the extinction time and $\{\pi(k)\}$ the invariant measure normalized so that $G(F(0)) = 1$ where $G(x) = \Sigma\pi(k)x^k$. Then for each integer $m \geq 1$, if $Z_0 = i$

$$u(j;m) = \lim_{n\to\infty} \Pr\,[Z_n=j\,|\,T=n+m]$$

determines a proper probability measure independent of i and moreover

$$\pi(j) = u(j;m)/(F_m^j-F_{m-1}^j).$$

Proof.

The conditional p.g.f. at time n is just

$$\frac{F_n(xF_m)^i - F_n(xF_{m-1})^i}{F_{n+m}^i - F_{n+m-1}^i}$$

and by the same argument as in the corollary of Theorem A, as $n \to \infty$ this behaves like $v\cdot(F_n(xF_m)-F_n(xF_{m-1}))/v\cdot(F_{n+m}-F_{n+m-1})$ whose limit, using (2b) and the previous theorem is just $G(xF_m)-G(xF_{m-1})$. As $x \to 1$ this difference approaches 1 which shows that $\{u(j;m)\}$ is a proper probability measure and that $u(j;m) = \pi(j)(F_m^j - F_{m-1}^j)$. □

1.5 INVARIANT MEASURES FOR THE IMMIGRATION PROCESS

In this section we are concerned with a process with immigration given by a p.g.f. $B(x)$. The equation of interest is

$$P(x) = B(x)\, P(F(x)), \qquad P(0) = 1, \tag{15}$$

since it was shown in [24] that an invariant measure is equivalent to a g.f. solution. F generates a non-supercritical process. In case the process is subcritical we allow the possibility that B is defective $(B(1) < 1)$ since, using suitable transformations, this framework subsumes both the supercritical process with proper immigration (if $q > 0$) and also to some extent invariant measures for the processes without immigration, although this was covered earlier, and more completely, directly from the Abel equation. We therefore content oursleves with stating the following theorems only for $\rho \le 1$. For simplicity of presentation we postulate $B(0) \ne 0$. As pointed out in [25, page 47] this results in no loss of generality.

<u>Theorem F.</u>

(a) If $B(1) = 1$ (15)possesses a unique g.f. solution given by

$$P(x) = \lim_{n\to\infty} \prod_{j=0}^{n} B(F_j(x))/B(F_j). \tag{16}$$

(b) If $B(1) = 1$ and $\rho < 1$ the (unique) solution has the property

$$P(1-su) = L(s) \tag{17}$$

where L is slowly varying at 0.

(c) If $B(1) < 1$ and $\rho < 1$ then all g.f. solutions are R-O varying*). There is a unique solution Q satisfying a regularly varying condition

$$Q(1-su) = s^{-\delta} L_1(s), \tag{18}$$

in which case δ is $\log B(1)/\log\rho$ and the solution given by

$$P(x)\ [1-A(x)]^{-\log B(1)/\log\rho} \tag{19}$$

where P is defined by (16) and A is the Yaglom p.g.f.

Remarks.

For the one-type case these results are explicitly or implicitly contained in [63],[70]. Except for the R-O variation assertion this theorem was enunciated and proven in [24]. The proof of (a) we give will be more direct. Also in (b) we give another uniqueness proof of (a) via regular variation which will lead into the analysis in (c).

Proof.

(a) Define $P_n(x) = \Pi_{j=0}^{n} B(F_j(x))/B(F_j)$. Since B is non-decreasing for each fixed x, the sequence $\{P_n(x)\}$ is non-decreasing and bounded above, since for example if $0 \leq x \leq F_k$ for some k then $P_n(x) \leq P_n(F_k) = \Pi_{j=0}^{n} B(F_{k+j})/B(F_j)$ and if n exceeds k this product telescopes leaving the upper bound $[B(0)]^{-k}$. The limit $P(x)$ as $n \to \infty$ therefore exists and is a g.f. by the continuity theorem. The identity $P_{n+1}(x) = P_n(F(x))B(x)/B(F_{n+1})$ and passage to the limit show that $P(x)$ satisfies (16) if $B(1) = 1$.

Turning to uniqueness let Q be any g.f. solution. By iteration $Q(x) = P_n(x)\ Q(F_{n+1}(x))/Q(F_{n+1})$. Let k be such that $0 \leq x \leq F_k$ producing the inequalities

* See Part 2 of this exposition.

$$1 \le Q(F_{n+1}(x))/Q(F_{n+1}) \le Q(F_{n+k+1})/Q(F_{n+1}) = [\prod_{j=0}^{k-1} B(F_{n+j+1})]^{-1} \to 1$$

as $n \to \infty$ showing by the last equation that $Q(x) = P(x)$.

(b) Recalling the remarks above, we temporarily ignore uniqueness, and suppose that $Q(x)$ is any g.f. solution. Anticipating the condition on Q at 1 we make the transformations $q(x) = Q(1-x)$, $f(x) = 1-F(1-x)$ and $b(x) = B(1-x)$ obtaining the corresponding equation $q(x) = b(x)q(f(x))$. From (0) if $\lambda \in [\rho,1]$ then $f(su) \le \lambda su \le su$ for small positive s whereby

$$1 \le q(\lambda su)/q(su) \le q(f(su))/q(su) = 1/b(su). \tag{20}$$

Let $s \to 0$ and slow variation obtains since $b(0) = B(1) = 1$.

We now use slow variation as an alternative to proving uniqueness. By iteration we obtain $q(x) = P_n(1-x)q(f_n(x))/q(f_n(1))$. Furthermore we may find a positive null sequence $\{\varepsilon_n\}$ such that

$$(1-\varepsilon_n)(1-A(1-x))v\cdot f_n(x)u \le f_n(x) \le (1+\varepsilon_n)(1-A(1-x))v\cdot f_n(x)u \tag{21}$$

and corresponding inequalities with 1 replacing x. By the uniform convergence property of slowly varying functions [43],[77] applied to (17) we see that $q(f_n(x))/q(f_n(1) \to 1$ giving $q(x) = \lim (n \to \infty) P_n(1-x) = P(1-x)$.

(c) From (20), Q being any solution, $s \ge 1$ and $1 \le \lambda \le \rho^{-1}$,

$$1 \le Q(1-(\lambda s)^{-1}u)/Q(1-s^{-1}u) \le 1/b(1)$$

which is R-O variation. One solution immediately available is that defined through (19) as direct substitution verifies, since from (a) the first factor satisfies (15) with $B(x)/B(1)$ in place of $B(x)$ and the second factor satisfies (by (4)) the same equation with $B(1)$ identically replacing $B(x)$. Moreover by (17) and since $1-A(1-su) = sL_2(s)$, L_2 slowly varying, we see that (18) is obeyed with $L_1(s) = L(s)L_2(s)$.

Next suppose in addition that Q satisfies (18) for some δ.

Setting $q(x) = Q(1-x)$ and iterating we find that $q(x) = P_n(1-x) \cdot q(f_n(x))/q(f_n(1))$ and using (21) once more, $\lim(n\to\infty) q(f_n(x))/q(f_n(1)) - [1-A(1-x)]^\delta$, which limit must satisfy $[1-A(x)]^\delta = B(1) \cdot [1-A(F(x))]^\delta$ and comparison of this equation with (4) shows that $\delta = \log B(1)/\log \rho$. This gives a representation for Q identical with (19) and ends the proof. □

Corollary.

With regard to (11) when $\rho < 1$ there is a unique solution satisfying, for some real c,

$$\lim_{s\to 0} G(1-\lambda su) - G(1-su) = c\log\lambda, \quad \lambda > 0$$

and then c must equal $(\log\rho)^{-1}$ and $G(x) = \log(1-A(x))/\log \rho$.

Proof.

Transform (11) into (15) via $P(x) = \exp G(x)$ and $B(x) = e^{-1}$ and then we use (18) and (19). □

The construction in (c) shows that solutions may be produced as follows. Let $G(x)$ be the g.f. of any invariant measure for the process without immigration. Then $P(x) \exp[-G(x)\log B(1)]$ solves (15), and the choice $G(x) = \log(1-A(x))/\log \rho$ singles out the regularly varying solution.

Having exhibited the existence of solutions which can be decomposed as the product of two generating functions, one resulting from proper immigration and the other from an (defective) immigration whose p.g.f. is identically constant, we now show this to be the general case.

Theorem G.

(a) A g.f. Q is a solution to (15) iff it is of the form

$$Q(x) = P(x)R(x)$$

where $R(x)$ is a g.f. satisfying

$$R(x) = B(1)R(F(x)), \qquad R(0) = 1.$$

(b) Furthermore if $\rho < 1$ then R is a g.f. solution if and only if

$$R(x) = H(A(x))$$

where H is a g.f. satisfying the linearized equation

$$H(s) = B(1)H(1-\rho+\rho s), \qquad H(0) = 1.$$

Proof.

(a) Clearly any function of the specified form satisfies (15). Conversely notice that if Q is a solution then

$$Q(x) = [\prod_{j=0}^{n-1} B(F_j(x))Q(F_n(x))] = P_{n-1}(x)[Q(F_n(x))\prod_{j=0}^{n-1} B(F_j)].$$

Letting $n \to \infty$ manifests the necessity of the representation and $R(x) = \lim\ (n\to\infty)\ Q(F_n(x))\prod_{j=0}^{n-1} B(F_j)$.

(b) The details are similar to previous lemmas. □

Remarks.

The immigration distribution enters into an invariant measure only through the slowly varying component $P(x)$ and the value $B(1)$ which are independent of any invariant measure chosen. In view of the fact that P varies slowly and A varies regularly, it is clear that the assertions of theorem F(c) arise only through the function H given in Theorem G(b). Although the case $\rho = 1$, $B(1) < 1$ is not of probabilistic significance it can be deduced on the basis of Theorem G(a) that (15) possesses a unique g.f. solution in this situation. We do not pursue this and merely note in passing that to prove this fact, there is no need (see [74] page 416) to involve a result of Rozmus-Chmura [58], since it suffices to handle $B(x) \equiv$ constant.

Quine [56] in case $\rho < 1$ has examined the limiting behaviour of the immigration process. He has shown that the transition probabilities tend to a proper limit iff the expectation of the logarithm of the immigration random vector is finite and then the p.g.f. of the limit satisfies (16). Therefore with regard to theorem F(b), $L(0+) < \infty$, iff this condition holds. Kaplan [35] has looked at the similar problem for $\rho = 1$.

1.6 THE SUPERCRITICAL PROCESS AND THE POINCARÉ EQUATION

This final section reviews some aspects of the normed supercritical process $(1 < \rho < \infty)$ bearing on characterizations of the limit.

It was shown in [26] that there exists a sequence $\{x_n\}$, called backward iterates, no member of which equals q or 1 such that

$$F(x_{n+1}) = x_n, \qquad n=0,1,2,\ldots . \tag{22}$$

If $c_n = -\log x_n$ (componentwise) one deduces from this equation that $\{\exp(-c_n \cdot Z_n)\}$ is a martingale adapted to $\{Z_n\}$. By writing (22) as $1-x_n = (M-E(x_{n+1}))(1-x_{n+1})$ and using (0), arguments of the sort giving (1a) and (1b) also yield

$$\lim_{n\to\infty} (1-x_n)/v\cdot(1-x_n) = u, \tag{23}$$

and

$$\lim_{n\to\infty} v\cdot(1-x_n)/v\cdot(1-x_{n+1}) = \rho. \tag{24}$$

A Taylor expansion shows that $\{c_n\}$ has the same asymptotic properties as $\{1-x_n\}$, displayed by (23) and (24), on which basis

$$\phi(s) = \lim_{n\to\infty} F_n(\exp(-sc_n)) \tag{25}$$

exists and satisfies the Poincaré equation

$$\phi(\rho s) = F(\phi(s)). \tag{26}$$

The ith component of ϕ is just the Laplace Stieljes transform of the limit random variable $W^{(i)} = \lim(n\to\infty)\ c_n \cdot Z_n$ given that $Z_0 = e_i$. From (26), it is apparent that both $\phi(0+)$ and $\phi(\infty)$ are fixed points of F which cannot be equal because $\phi_n(1) = F_n(\exp(-c_n))$

$\equiv x_0$ which is not a fixed point. By monotonicity $\phi(0+) = 1$ and $\phi(\infty) = q$ showing that the limit is a.s. finite and moreover is zero a.s. only on the set of extinction (cf. [75]).

If one defines $\gamma_n = v \cdot c_n$ then because $c_n \sim \gamma_n u$ it is also true that given $Z_0 = e_i$, $\lim(n\to\infty)\gamma_n Z_n \cdot u = W^{(i)}$ a.s. It can also be shown [28] that on the set of non extinction $Z_n/Z_n \cdot u \to v$ a.s. as $n \to \infty$, producing the final result

$$\lim_{n\to\infty} \gamma_n Z_n = W^{(i)} v \quad \text{a.s.}.$$

(In [26] convergence in probability was obtained.) Our final theorem characterizes solutions of (26) and relates the constants $\{\gamma_n\}$ to the classical norming by $\{\rho^{-n}\}$.

Theorem H.

Up to a scale factor there is a unique, strictly decreasing and convex solution to (26) satisfying $\phi(0+) = 1$.

$$v \cdot (1-\phi(s)) = sL(s) \tag{27}$$

where L is slowly varying at 0.

$$\gamma_n \sim \rho^{-n} L(\rho^{-n}) \quad \text{as } n \to \infty. \tag{28}$$

$$\int_0^s P[W^{(i)} > t]dt \sim u_i L(s^{-1}) \quad \text{as } s \to \infty. \tag{29}$$

Remark.

This appeared as Theorems 2.2 and 2.3 of [26], following one-type results in [74]. The proof below of regular variation is simpler and along the lines of [76].

Proof.

Let $\psi(s)$ denote any appropriate solution. The function $r(s) = v \cdot (1-\psi(s))/s$ therefore decreases as s increases. For $1 \le \lambda \le \rho$, $1 \equiv r(s)/r(s) \ge r(\lambda s)/r(s) \ge r(\rho s)/r(s) \equiv v \cdot (1-F(\phi(s)))/\rho v \cdot (1-\phi(s))$ and because $\phi(0+) = 1$, the right hand inequality approaches 1 as $s \to 0$ proving (27).

The sequence $\{\psi(s\rho^{-n})\}$ for any fixed $s > 0$ satisfies (22) and if used to construct normalizing constants will also produce a proper limit. Since we have a strong limit, non zero on a set of positive measure, $\lim(n\to\infty)\ v\cdot(1-\psi(s\rho^{-n}))/v\cdot(1-x_n) = K(s)$ where $K(s)$ is a finite function non-zero for $s > 0$. From (27), ψ in place of ϕ, we obtain $K(\rho s) = \rho K(s)$ and because K is a concave function $K(s) = Ks$ for some constant $0 < K < \infty$. Asymptotically therefore taking into account the directional convergence of norming vectors, $-\log \psi(s\rho^{-n}) \sim Ksc_n$. For $\varepsilon > 0$ and n sufficiently large $(1-\varepsilon)Ksc_n \le -\log\psi(s\rho^{-n}) \le (1+\varepsilon)Ksc_n$, and applying F_n and passing to the limit gives $\psi(s) = \phi(Ks)$ the uniqueness assertion.

From the above construction $x_n = \phi(\rho^{-n})$, and (28) follows directly from (27). It can also be shown that $1-\phi(s) \sim v\cdot(1-\phi(s))u$ as $s \to 0$ and Karamata's Tauberian theorem gives (29), u_i being ith component of u. □

In view of Kesten and Stigum [40], $L(0+) < \infty$ iff ZlogZ holds in which case γ_n is asymptotically ρ^{-n} up to a multiplicative constant and $E[W^{(i)}] < \infty$. In general

$$E[(W^{(i)})^{\tau}] < \infty \qquad \text{for} \qquad 0 \le \tau < 1$$

and

$$P[W^{(i)} > s] = o(s^{-1}L(s^{-1})), \quad s \to \infty.$$

These results may be compared with their counterparts following Theorem A. In one dimension the Poincaré equation inverts to the Schröder equation and this resemblance appears naturally from this viewpoint (see [74] and [76]).

As a final note, we remark that, in line with the analytical arguments of this paper, we could obtain convergence in distribution (a weaker result) by dispensing with the probabilistic tool - the martingale convergence theorem - which was used to exhibit (25) as a solution to (26). This idea could in turn be extended to study (26) in the more general setting than when F is a p.g.f., thus "reversing" the (one-type) arguments in [62].

PART 2

2.1 THE ROLE OF R-O VARIATION

Definition. A function $\mathcal{H}$ is said to be R-O varying at infinity if it is real-valued, positive and measurable on $[A,\infty)$ for some $A > 0$, and for all $x \geq A$,

$$M_1 \leq \mathcal{H}(\lambda x)/\mathcal{H}(x) \leq M_2, \qquad 1 \leq \lambda \leq a \tag{30}$$

where M_1, M_2 and a are any constants satisfying $0 < M_1 \leq 1$, $1 \leq M_2 < \infty$, $1 < a < \infty$.

Clearly any function regularly varying at infinity is also R-O varying there, so the notion is a generalization of the concept of regular variation. A simple example of an R-O varying function which is not regularly varying is $\mathcal{H}(x) = x^{\alpha}\{1+|\sin x|\}$; indeed, any function $\mathcal{H}(x)$ positive and measurable on $[A,\infty)$ for $A > 0$ and satisfying

$$0 < A_1 < x^{-\alpha}\mathcal{H}(x) < A_2 < \infty \tag{31}$$

for constants A_1, A_2, α is clearly R-O varying. More generally still, if $\mathcal{H}_1$ and $\mathcal{H}_2$ are positive and measurable on $[A,\infty)$, $\mathcal{H}_1$ is R-O varying and $\mathcal{H}_1(x) \asymp \mathcal{H}_2(x)$, by which we mean, for $x \geq A$,

$$0 < A_1 < \mathcal{H}_2(x)/\mathcal{H}_1(x) < A_2 < \infty, \tag{32}$$

then $\mathcal{H}_2$ is R-O varying. The notion of R-O variation was introduced by J.Karamata [36], [37], who was motivated by work of V.G. Avakumović [2],[3]. In the probabilistic literature it occurs without mention of such antecedents, in the work of Feller beginning in 1963, who considers it only for monotone functions and eventually calls it dominated variation (see also section 2.5). An account of the theory of R-O variation, together with historical notes and references, may be found in [77].

One striking property of R-O varying functions, the one with which we shall be most concerned, is that there exist $\alpha \geq 0$, $\beta \geq 0$ such that

$$\left.\begin{array}{l} y^{-\alpha}\mathcal{H}(y) \le M_2 x^{-\alpha}\mathcal{H}(x) \\ M_1 x^{\beta}\mathcal{H}(x) \le y^{\beta}\mathcal{H}(y) \end{array} \quad y \ge x \geq A, \right\} \tag{33}$$

$$\alpha = \log M_2/\log a, \qquad \beta = -\log M_1/\log a, \tag{34}$$

that is, $x^{-\alpha}\mathcal{H}(x)$ is "almost decreasing" and $x^{\beta}\mathcal{H}(x)$ is "almost increasing", in the terminology of S.N.Bernstein. In fact, this property is essentially equivalent to R-O variation (e.g. [77, Theorem A.2]). Clearly, in case of monotone $\mathcal{H}$, with which we shall generally have to deal, there is a considerable simplification in that one of M_1 and M_2 is unity.

Just as much of the interest in regular variation in a probabilistic context focuses on Tauberian theorems, such as Karamata's Tauberian Theorem, such questions are also of interest in connection with the generalized concept. Insofar as the properties of R-O variation are in general not nearly as elegant as those for ordinary regular variation, one may apparently say relatively little in this direction. Let $U(t)$ be a monotone non-decreasing function on $[0, \infty)$; then

$$e^{-1}\int_0^{1/x} U(dt) \le \int_0^{1/x} e^{-xt}U(dt) \le \int_0^{\infty} e^{-xt}U(dt). \tag{35}$$

In the special case where $U(dt) = u(t)dt$ for a function $u(t) \ge 0$, $\not\equiv 0$ non-increasing on $(0,\infty)$, also

$$\begin{aligned} \int_0^{\infty} e^{-xt}U(dt) &= \int_0^{\infty} e^{-xt}u(t)dt \le \int_0^{1/x} u(t)dt + u(1/x)\int_{1/x}^{\infty} e^{-xt}dt \\ &= \int_0^{1/x} u(t)dt + (1/x)u(1/x) \le 2\int_0^{1/x} u(t)dt, \end{aligned} \tag{36}$$

so in this special case

$$\int_0^x u(t)dt \asymp \int_0^{\infty} e^{-t/x}u(t)\, dt, \qquad x > 0 \tag{37}$$

so that if either is R-O varying as $x \to \infty$, so is the other by (32).

These kinds of properties are mentioned a number of times by Feller, but are made more or less explicit only in [34, section 6]. Clearly (37) promises to be useful in connection with right-tail behaviour of a probability distribution function $\mathfrak{F}(t)$ on $[0,\infty)$, with $u(t) = 1-\mathfrak{F}(t)$; for left-tail behaviour (near the origin) of such a function as deduced from its Laplace transform, (35) may have to suffice, with $U(t) = \mathfrak{F}(t)$. It should be mentioned that both Avakumović and Karamata were interested in generalizing regular variation precisely in connection with a theorem of Tauberian type, which, however, does not seem to us particularly useful in probabilistic contexts.

In the theory of simple branching processes R-O variation, like regular variation, arises in a natural way (without artificial assumptions) since both are induced by the essential structure of the processes as manifestations of functional iteration, and, more specifically, by the associated functional equations of Abel-Schröder type which permeate the topic. The duality mentioned earlier occurs in the sense that if a limiting distribution has a regularly varying property in either tail, it seems to have at least a R-O varying property in the other tail. Even though the R-O property is a much weaker one, it is still associated, in consequence of (33), with an asymptotic power-law structure.

The ensuing discussion to illustrate these points consists of a number of more or less graphic, if disjoint, examples, the first of which is a discussion of the limit law for the ordinary supercritical case, which casts some light even on a question raised by Harris in 1948.

2.2 THE SUPERCRITICAL CASE

Consider the supercritical Galton-Watson process $\{Z_n\}$, $(Z_0=1)$, with offspring probability generating function $F(s) = \Sigma_{j=0}^{\infty} p_j s^j$, $p_j \neq 1$, all j, $1 < m = EZ_1 < \infty$. Let W be the well-known limit random variable arising in this situation (e.g. [1] though we tend to follow the approach of [62]). Let $\phi(s) = E(e^{-sW})$, $K(s) = -\log\phi(s)$, $k(s) = -\log F(e^{-s})$, $s \geq 0$: then

$$K(ms) \quad = \quad k(K(s)), \qquad s \geq 0$$

from which, it follows that for $1 \leq \lambda \leq m$, $s \geq 0$

$$1 \;\geq\; \frac{K(\lambda s)/\lambda s}{K(s)/s} \;\geq\; \frac{K(ms)/ms}{K(s)/s} \;=\; \frac{k(K(s))}{mK(s)} \;\geq\; \frac{1}{m}\lim_{s\to\infty}\frac{k(s)}{s}. \qquad (38)$$

The equality follows from the functional equation, and the first two inequalities from the concavity of K on $[0,\infty)$ and the fact that $K(0) = 0$. Letting $s \to 0^+$ in these first three relationships, bearing in mind that $k(s)/s \to m$, yield that $xK(1/x) = L(x)$, a slowly varying function at infinity. Using $K(1/x) \sim (1-\phi(1/x))$ leads ultimately to the information contained in Theorem H of Part 1.

The last inequality in (38) follows in virtue of the concavity of k on $[0,\infty)$; and $r = \lim k(s)/s$ is readily identifiable as the first point of increase of the distribution of $\{Z_1\}$. By a device viz. transformation from the p.g.f. $F(s)$ to the p.g.f.

$$[F(q+(1-q)s)-q]/(1-q)$$

where q is the extinction probability, we may assume without loss of generality in the future that $P[Z_1 = 0] = 0 = P[W = 0]$, so that r is an integer ≥ 1. We then see from (38) that $\mathcal{H}(s) = K(s)/s$ satisfies the definition of R-O variation, with $a = m$, $M_1 = r/m$, $M_2 = 1$; and this gives us some information about the distributions of W at the origin. Thus the second of the inequalities (33), with $x=1$, yields the right-hand side of

$$e^{-1}P[W \leq 1/s] \;\leq\; \phi(s) \;\leq\; \exp\{-c(r/m)s^{\delta}\}, \qquad s \geq 1 \qquad (39)$$

where $c = \mathcal{H}(1)$, $\delta = \log r/\log m$, the left-hand side following from (35). Naturally (39) is interesting only if $r > 1$, and then bears out a remark of Harris [20] (Remark c below Theorem 3.3).

In the remaining case, which we consider henceforth in this section, $r=1$, which amounts to $F'(q) > 0$ in the untransformed process, on the other hand, we have a great deal of information already from the work of Dubuc [13] who has shown that the strictly positive continuous density of W on $(0,\infty)$, $w(t)$, in this case satisfies

$$c_1 t^{\gamma-1} \leq w(t) \leq c_2 t^{\gamma-1}, \qquad t \in (0,1] \tag{40}$$

where $\gamma = -\log p_1/\log m$, and c_1 and c_2 are positive constants. In particular $w(1/x)$, $x \geq 1$ is R-O varying, by (31). Dubuc shows that in general it is not true that $t^{1-\gamma}w(t)$ approaches a limit as $t \to 0+$; this does not, however, necessarily preclude $w(1/x)$ actually being regularly varying with index $1-\gamma$, for examples of slowly varying functions which continue to oscillate between two fixed limits are known [77, Exercise 1.12].

Further, Dubuc considers at length conditions equivalent to the existence of the limit $(t \to 0+)$ of $t^{1-\gamma}w(t)$; one such condition is the constancy of a certain continuous and positive function $M(\cdot)$ on $(0,\infty)$, which also satisfies $M(ms) = M(s)$. The constancy of this function, introduced by Harris [20], has been the focus of much study, especially by Karlin and McGregor [39]. Of interest to us here is Harris' theorem 4.2:

$$\lim_{u\to 0} \frac{D(mu)}{D(u)} = \exp\left\{\frac{1}{\Gamma(\gamma+1)} \int_1^m \frac{M(v)}{v}\, dv\right\} \tag{41}$$

where

$$D(u) = \exp\{-\int_u^1 (P[W\leq v]/v^{\gamma+1})dv\}.$$

(Harris' proof is given under the condition $EZ_1^2 < \infty$, but is easily adjusted to omit this condition.)

Now, if in (41) the result continued to hold with m replaced by any λ in some subset of positive measure of $[1,\infty)$, then $D(1/x)$ would be regularly varying, and, it would follow from the Characterization Theorem of such functions [77, Theorem 1.3] that for all $\lambda > 0$ and some constant $-\infty < \tau < \infty$

$$\lambda^{\tau} = \exp\{\frac{1}{\Gamma(\gamma+1)} \int_1^{\lambda} \frac{M(v)}{v} dv\}$$

whence $M(x)$ = constant for all $x > 0$.

This speculation aside, it is nevertheless clearly true that for $0 < u \leq 1$, $1 \leq \lambda \leq m$

$$1 \leq D(\lambda u)/D(u) \leq 1/M_1$$

for some $M_1 > 0$, in view of (40), whence $D(1/x)$ clearly satisfies the definition of R-O variation with $\alpha = m$, $M_1 = 1$. (This again provides information about the distribution of W near the origin, but not comparable to that given by (40)).

The preceding remarks are related to two further theorems of Harris [20], viz. Theorems 3.3 and 3.4 which both involve functions which are "almost" regularly varying. We can use functional-equation techniques, associated with R-O variation, to cast light on both, although we shall not do so in detail. Thus we may obtain

$$\phi(s) \geq \text{const.}\ s^{-\gamma}, \qquad s \geq 1,$$

where γ is as in (12), and if $F(s)$ is actually a polynomial of degree d, then

$$H(t) \equiv -K(-t) \leq \text{const.}\ t^{\rho}, \quad t \geq 1,$$

where $\rho = \log d/\log m$. Note that the indices γ and ρ, emanating from (33), are as in Harris' theorems. Here and in the sequel, "const." refers to positive finite numbers which can be evaluated (in principle) in the manner leading to (39).

2.3 OTHER LIMIT LAWS

(i) Process with Infinite Mean

In the theory of the simple Galton-Watson process with infinite mean ($m = \infty$) [7], [74], [31], [57] one considers a $U(t)$ defined, continuous and strictly increasing on $[1,\infty)$ with $U(1) = 0$, $U(\infty) = \infty$, and slowly varying at infinity. Denote by $W(t)$ the inverse function, defined on $[0,\infty)$ of $U(t)$, and assume

$$\phi(t) \equiv U\{(1-q)/(1-F(1-[(1-q)/W(t)]))\}$$

$t \geq 0$, satisfies the conditions:

(A) $\phi(t)$ is convex or concave on $[0,\infty)$;

(B) $c = \lim(t\to\infty)\ \phi(t)/t$ satisfies $0 < c < 1$.

The limit law which then arises has distribution function on $[0,\infty)$ given by

$$v(t) = 1 - [(1-q)/W(\Delta(t))], \qquad t > 0$$

where $\Delta(t)$ is continuous and strictly increasing from $\Delta(0+) = 0$ to ∞, and is convex or concave on $(0,\infty)$ as $\phi(t)$. Moreover, it satisfies the functional equation

$$\Delta(cx) = \phi(\Delta(x)), \qquad x \in (0,\infty).$$

Thus in the case of convexity of $\phi(t)$ (and so $\Delta(t)$), for $1 \leq \lambda \leq c^{-1}$, and $t > 0$

$$1 \geq \frac{\Delta(1/\lambda t)/(1/\lambda t)}{\Delta(1/t)/(1/t)} \geq \frac{\Delta(c/t)/(c/t)}{\Delta(1/t)/(1/t)} = \frac{\phi(\Delta(1/t))}{c\Delta(1/t)} \geq \frac{1}{c} \lim_{t\to 0^+} \frac{\phi(t)}{t}, \qquad (42)$$

the reasoning being analogous to that of (38):and the inequalities are reversed if ϕ is concave. Letting $t \to 0^+$ reveals, as is known, that $\Delta(x) = xL(x)$, where $L(x)$ is slowly varying at infinity.

We may also use (42) to make deductions about the behaviour of $\Delta(x)$ as $x \to 0^+$. In case of convexity, we need to assume $\phi'(0^+) = \lim(t\to 0^+)\ \phi(t)/t > 0$, in which case $t\Delta(1/t)$ is R-O varying, and we may deduce from the second of the inequalities (33) with $x=1$, that

$$\Delta(1/t) \geq \text{const. } t^{-\delta}, \qquad t \geq 1, \qquad (43)$$

where $\delta = \log \phi'(0^+)/\log c \geq 1$. The assumption $\phi'(0^+) > 0$ here is tantamount to the assumption $F'(q) > 0$ in the case $1 < m < \infty$, which on transformation to achieve $P[Z_1 = 0] = 0$, translates to $p_1 > 0$, in that case.

In case of concavity, since it is readily checked that $\phi'(0^+) \leq 1$ always, no similar assumption is necessary, and (43) holds with the inequality reversed (and with $\delta \le 1$).

If we permit immigration into the process [55],[31] according to a p.g.f. $B(s) = \Sigma_{j=0}^{\infty} b_j s^j$, $B(0) \ne 1$, then the proper limit law which results is concentrated on $(0,\infty)$ where it has continuous distribution function $p(t)$ satisfying

$$p(ct) = p(t)\, B(v(t)), \qquad t > 0$$

providing a priori that $\Sigma_{j=2}^{\infty} b_j \log U(j) < \infty$. Again operating on the functional equation we have for $1 \le \lambda \le c^{-1}$, $t > 0$

$$1 \ge \frac{p(1/t\lambda)}{p(1/t)} \ge \frac{p(1/tc^{-1})}{p(1/t)} = B(v(1/t)) \ge B(q)$$

so $p(1/t)$ is R-O varying at infinity, whence, in the now familiar manner

$$p(1/t) \gtrsim \text{const.}\ t^{-\beta}, \qquad t \ge 1$$

where $\beta = \log B(q)/\log c$. Thus we have some information on the behaviour of the limit distribution function near the origin; we are not aware of any investigation of the behaviour of its upper tail, $1-p(t)$, for large t.

(ii) Subcritical Jiřina Process

The simple kinds of discrete-time continuous-state processes considered in [80],[81] are governed by a non-degenerate cumulant generating function (c.g.f.) $k(s)$, $s \ge 0$, which describes the quantity of offspring per unit quantity of parent, and so is required to be infinitely divisible; put m for the mean of the corresponding distribution. When $m < 1$ a phenomenon essentially foreign to the ordinary discrete (Galton-Watson) process occurs if the first point of increase, r, of the offspring distribution satisfies $r > 0$. There is then an a.s. limit random variable for the normed process, which is proper and non-degenerate, and whose c.g.f. $\gamma(s) = -\log\phi(s)$, $s \ge 0$, satisfies the functional equation

$$\gamma(rs) = k(\gamma(s)), \qquad s \in [0,\infty).$$

Thus, for $1 \le \lambda \le r^{-1}$, $t > 0$

$$1 \le \frac{\gamma(1/\lambda t)/(1/\lambda t)}{\gamma(1/t)/(1/t)} \le \frac{\gamma(1/r^{-1}t)/(1/r^{-1}t)}{\gamma(1/t)/(1/t)} = \frac{k(\gamma(1/t))/(1/r^{-1}t)}{\gamma(1/t)/(1/t)} \le \frac{m}{r}$$

and again we may conclude that

$$\gamma(x) = xL(x)$$

where $L(x)$ is slowly varying at infinity; and $x\gamma(1/x)$ is R-O varying at infinity. From this last, since as $x \to \infty$, $x\gamma(1/x) \sim x(1-\phi(1/x))$, it follows that

$$\int_0^\infty e^{-t/x}\{1-\mathcal{F}(t)\} \, dt$$

is R-O varying at infinity, whence by (37), so is

$$\int_0^x \{1-\mathcal{F}(t)\} \, dt$$

where $\mathcal{F}(t)$ is the distribution function of the limiting random variable. These results yield that (by (35))

$$\mathcal{F}(1/x) \le \text{const. } e^{-xL(x)};$$

and

$$\int_0^x \{1-\mathcal{F}(t)\}dt \le \text{const. } x^{\alpha}, \quad x \ge 1$$

for some $\alpha > 0$.

2.4 INVARIANT MEASURES

For a non-critical Galton-Watson process, without immigration (where $0 < F(0) < 1$), or with immigration (where, additionally, $0 < B(0) < 1$), invariant measures may be studied in unified manner, by examining the problem merely for the subcritical process $(m < 1)$ with possibly defective immigration distribution $(B(1-) \le 1)$, as noted already in our section 1.5, and in [70]. For such a process denote an invariant measure by $\{a_r\}$, $r \ge 1$ $(\{a_r\} \ge \{0\}, \ne \{0\})$, and its generating function by $P(s) = \sum_{r=1}^{\infty} a_r s^r$, $s \in [0,1)$. On the basis of the proof of Theorem F(c), we obtain thus

$$r(x) \equiv P(1-x^{-1})$$

is <u>always</u> R-O varying at infinity, possible non-uniqueness of invariant measure notwithstanding. It is this always true for this particular process that

$$P(1-x^{-1}) \leq (1/B(0))x^{\alpha}, \qquad \alpha \geq 1 \tag{44}$$

with $\alpha = \log B(0)/\log m$, which implies by an argument analogous to (35) that

$$\sum_{r=1}^{n} a_r \leq \text{const.}\ n^{\alpha}.$$

This result is, as one might expect, rather weak. If we primarily consider a subcritical ordinary Galton-Watson process (without immigration), and write $G(s)$ for the generating function of any one of its invariant measures, with $G(F(0)) = 1$, then this translates to the above framework with the same $F(s)$, and $B(s) = e^{-1}$; and

$$P(1-x^{-1}) = \exp G(1-x^{-1}), \qquad x \geq 1.$$

Thus here $\alpha = -1/\log m$, and we obtain from (44)

$$G(1-x^{-1}) \leq 1 - (\log x/\log m), \qquad x \geq 1.$$

We see from Theorem D(b), however, that actually

$$G(1-x^{-1}) \sim -\log x/\log m$$

as $x \to \infty$.

2.5 THE CRITICAL CASE

Hitherto in connection with the Galton-Watson process, our discussion of R-O variation has been confined to the non-critical situations $(m \equiv F'(1-) \neq 1)$. The notion also enters into the theory of the critical case, where $m = 1$; this theory is well-known if $0 < \sigma^2 \equiv F''(1-) < \infty$, so we suppose henceforth that $\sigma^2 = \infty$, and merely report and discuss some results of Kallenberg [34] in this connection.

Kallenberg is influenced by the "dominated variation" approach of Feller, and it is necessary to give, at this stage, that version of the R-O definition.

Definition. A positive monotone function $\mathcal{H}$ on $[A,\infty)$, $A > 0$, is said to vary dominatedly (at infinity) if for some fixed $\lambda_0 > 1$

$$\limsup_{x\to\infty} \mathcal{H}(\lambda_0 x)/\mathcal{H}(x) < \infty, \quad \text{if } \mathcal{H} \text{ is non-decreasing;} \tag{45a}$$

$$\limsup_{x\to\infty} \mathcal{H}(x/\lambda_0)/\mathcal{H}(x) < \infty, \quad \text{if } \mathcal{H} \text{ is non-increasing.} \tag{45b}$$

This form of definition makes it clear that for monotone functions one might also consider a related concept to R-O variation, where in (45a) and (45b) "lim sup" is replaced by "lim inf", and "$<\infty$" is replaced by ">1". Indeed this is done by Bari and Stechkin [4], who call dominated variation its condition (L_k), and the related concept the condition (L).

Returning to the critical Galton-Watson process, let us write $g(x) = 1-F'(1-x)$, and, if V is a random variable with the probability distribution generated by $F'(s)$, $Q(x) = P[V > x]$. Then the ordinary inverse function g^{-1} of g is well-defined in the right neighbourhood of the origin; and for $0 \le u \le 1$, we may define $Q^{-1}(u) = \inf\{x; Q(x) \le u\}$. Consider now the two conditions:

(A_1) $\quad \limsup_{x\to\infty} Q(x/2)/Q(x) < \infty$;

(A_2) $\quad \liminf_{x\to\infty} g(1/(2x))/g(1/x) > 1$.

Insofar as $Q(x)$ is decreasing as x increases, as is $g(1/x)$, we see (A_1) and (A_2) are respectively (45b) and its "lim inf" version. Focus attention now on the distribution of the time to extinction T, and put $\alpha(n) = P[T > n]$. Then Kallenberg shows, in the results stated, that:

$(A_2) \Rightarrow \alpha(n) \asymp g^{-1}(1/n)$, (his Lemma 4.7);

$$(A_1) \text{ and } (A_2) \Rightarrow \alpha(n) \asymp 1/Q^{-1}(1/n), \qquad \text{(his Theorem 6.1)}.$$

Perhaps we should mention in this context the now well-known result that

$$1 - F(1 - x) - x = -x\Lambda(x) \Rightarrow \Lambda(\alpha(n)) \sim 1/\tau n,$$

as $n \to \infty$ where $\Lambda(x) = x^{\tau}L(1/x)$, $0 < \tau \le 1$, L being slowly varying at infinity. The result due to Karamata [38] was rediscovered by Slack [84].

In a later paper, Slack [85, Lemma 1] shows that

$$\liminf_{n\to\infty} n\{1-(\alpha(n+1)/\alpha(n))\} \ge 1, \qquad \limsup_{n\to\infty} \alpha(kn)/\alpha(n) \le k^{-1},$$

k a positive integer always hold true in the case $m = 1$. Clearly these are also both manifestations of an R-O type property of a non-increasing sequence $\{\alpha(n)\}$ (c.f. [7]) which arises completely naturally in this branching process context.

BIBLIOGRAPHY

1. K.B. Athreya and P.E. Ney, Branching Processes, Springer-Verlag, New York (1972).
2. V.G. Avakumović, C.R. Acad. Sci., Paris, 200:1515 (1935).
3. V.G. Avakumović, Rad Jugoslovenske Akad. Znatnosti i Umjetnosti, 254:167 (1936).
4. N.K. Bari and S.B. Stechkin, Trudy Mosk. Mat. Obschestva, 5: 483 (1956).
5. R. Bojanić and E. Seneta, J. Math. Anal. Applic., 34:302 (1971).
6. N.G. de Bruijn, Nieuw Arch. Wisk., 7:20 (1959).
7. H. Cohn, Z. Wahrscheinlichkeitsth., 38:73 (1977).
8. R.R. Coifman and M. Kuczma, Aequationes Math., 2:332 (1969).
9. S. Dubuc, Proc. Amer. Math. Soc., 21:324 (1969).
10. S. Dubuc, Studia Math., 34:69 (1970).
11. S. Dubuc, Ann. Inst. Fourier (Grenoble), 21:171 (1971).

12. S. Dubuc, Séminaire d'Analyse Moderne No.7, Dept. Math. Université de Sherbrooke, Canada (1971).

13. S. Dubuc, Z. Wahrscheinlichkeitsth., 19:281 (1971).

14. S. Dubuc and E. Seneta, Ann. Prob., 4:490 (1976).

15. W. Feller, Arch. Math., 14:317 (1963).

16. W. Feller, An Introduction to Probability Theory and Its Applications, 2nd ed., Wiley, New York (1971).

17. J. Galambos and E. Seneta, Proc. Amer. Math. Soc. 41:110 (1973)

18. M.I. Goldstein and F.M. Hoppe, J.Math.Anal.Applic. (to appear).

19. M.I. Goldstein and F.M. Hoppe, J. Multivariate Anal. (to appear).

20. T.E. Harris, Ann.Math.Statist., 19:474 (1948).

21. T.E. Harris, The Theory of Branching Processes, Springer, Berlin (1963).

22. C.R. Heathcote, E.Seneta and D.Vere-Jones, Teor. Veroiatn. Primenen., 12:341 (1967).

23. C.C. Heyde, Ann. Math. Statist., 41:739 (1970).

24. F.M. Hoppe, J. Appl.Prob., 12:219 (1975).

25. F.M. Hoppe, Functional Equations with Applications to Multitype Galton-Watson Branching Processes, Ph.D. Dissertation, Princeton University (1975).

26. F.M. Hoppe, Ann. Prob., 4:393 (1976).

27. F.M. Hoppe, Ann. Prob., 5:291 (1977).

28. F.M. Hoppe, Almost-sure convergence of supercritical multitype branching processes without Z log Z moments. Unpublished report, University of Alberta (1977).

29. F.M. Hoppe, Stoch. Proc. Applic., 5:57 (1977).

30. F.M. Hoppe, Proc. Amer.Math.Soc., 64:326 (1977).

31. I.L.Hudson and E.Seneta, J. Appl.Prob., 15: to appear (1977).

32. P. Jagers, Branching Processes with Biological Applications, J. Wiley, London (1975).

33. A.Joffe and F.Spitzer, J.Math.Anal.Appl., 19:409 (1967).

34. O. Kallenberg, Math. Nachr., 77 :7 (1977).

35. N. Kaplan, Ann. Prob., 1:947 (1973).

36. J. Karamata, Bulletin Intern. de l'Acad. Yougoslave, Zagreb 29 et 30:117 (1935).
37. J. Karamata, Rad Jugoslovensko Akademije Znatnosti etnosti, 254:187 (1936).
38. J. Karamata, Recueil des Travaux de l'Academie Serbe des Sciences 35, Institut Mathematique, 3:45 (1953).
39. S. Karlin and J. McGregor, Trans.Amer.Math.Soc., 132:115 (1968).
40. H. Kesten and B.P. Stigum, Ann.Math.Statist., 37:1211 (1966).
41. E. Khalili-Francon, Séminaire de Probabilites VII 1971-72, Springer Lecture Notes in Mathematics 321:122 and 465:589 (1973)
42. A. Khinchine, Giorn.Inst.Ital.Attuari, 7:365 (1936).
43. J. Korevaar, T. van Aardenne-Ehrenfest and N.G. de Bruijn, Nieuw Arch. Wisk., 23:77 (1949).
44. M. Kuczma, Mathematica (Cluj), 3:79 (1961).
45. M. Kuczma, J. Austral.Math.Soc., 4:149 (1964).
46. M. Kuczma, Bull.Acad.Polon.Sci., 13:645 (1965).
47. M. Kuczma, Mathematica (Cluj), 8:279 (1966).
48. M. Kuczma, Mathematica (Cluj), 9:285 (1967).
49. M. Kuczma (ed.), Równania Funkcyjne w Teorii Procesów Stochastycznych. Wydawnictwo Uniwersytetu Śląskiego 47, Katowice (1972).
50. M. Kuczma, Wiadomości Matematyczne (Warsaw), 16:13 (1973).
51. M. Kuczma, J. Austral.Math.Soc. Ser.A, 22:135 (1976).
52. C. Lipow, Two Branching Models with Generating Functions Dependent on Population Size, Ph.D. Thesis, University of Wisconsin (1971).
53. C.J. Mode, Multitype Branching Processes, American Elsevier, New York (1971).
54. Y. Ogura and K. Shiotani, Osaka J. Math., 13:83 (1976).
55. A.G. Pakes, J. Appl.Prob., 13:17 (1976).
56. M.P. Quine, J. Appl.Prob., 7:411 (1970).
57. I.L. Romanow and E. Seneta, Bull.Inst.Math.Statist., 5:184 (1976).
58. M. Rozmus-Chmura, Publ.Math., 15:45 (1968).

59. H. Rubin and D. Vere-Jones, J. Appl. Prob., 5:216 (1968).
60. T.H. Savits, Ann.Prob., 3:61 (1967).
61. E. Seneta, J. Appl. Prob., 4:489 (1967).
62. E. Seneta, Ann.Math.Statist., 39:2098 (1968).
63. E. Seneta, Adv.Appl.Prob., 1:1 (1969).
64. E. Seneta, J. Austral.Math.Soc., 10:207 (1969).
65. E. Seneta, Math. Biosciences, 6:305 (1970).
66. E. Seneta, Math. Biosciences, 7:9 (1970).
67. E. Seneta, J. Roy.Statist.Soc. Ser. B, 32:149 (1970).
68. E. Seneta, Bull. Austral. Math. Soc., 2:359 (1970).
69. E. Seneta, Nature (London), 225:766 (1970).
70. E. Seneta, J. Appl.Prob., 8:43 (1971).
71. E. Seneta, J. Appl.Prob., 10:206 (1973).
72. E. Seneta, Ann.Prob., 1:1057 (1973).
73. E. Seneta, Non-Negative Matrices, Wiley, New York (1973).
74. E. Seneta, Adv.Appl.Prob. 6:408 (1974).
75. E. Seneta, Stoch.Proc.Applic. 3:35 (1975).
76. E. Seneta, Statistical Distributions in Scientific Work (G.P.Patil, S.Kotz and J.K.Ord, eds.), 3:249, Reidel, Dordrecht (1975).
77. E. Seneta, Regularly Varying Functions, Springer Lecture Notes in Mathematics 508, Berlin (1976).
78. E. Seneta, Aequationes Math., 14:457 (1976).
79. E. Seneta and D. Vere-Jones, J. Appl.Prob., 3:403 (1966).
80. E. Seneta and D. Vere-Jones, Z.Wahrscheinlichkeitsth., 10:212 (1968).
81. E. Seneta and D. Vere-Jones, Czechoslovak Math.J., 19:277 (1969)
82. B.A. Sevastyanov, Itogi Nauki, Seria Matematika, Moscow 1968:5 (1967).
83. B.A. Sevastyanov, Vetviashchiesia Protsessi, Nauka, Moscow (1971).
84. R.S. Slack, Z. Wahrscheinlichkeitsth., 9:139 (1968).
85. R.S. Slack, Z. Wahrscheinlichkeitsth., 25:31 (1972).
86. A. Spataru, Rev.Roum.Math.Pures Appl., 21:767 (1976).

87. F. Spitzer, Symposium on Probability Methods in Analysis, Springer Lecture Notes in Mathematics 31:296, Berlin (1967).
88. V.V.Tănase, Studii. Cerc. Mat., 25:1007 (1973).
89. V.A. Vatutin, Matem.Sbornik., 103:253 (1977).
90. A.M. Zubkov, Teor.Veroiatn.Primenen, 20:614 (in Russian) and Theory Probab.Appl., 20:602 (translation) (1975).

REMARKS ON THE STRUCTURE OF TREES WITH APPLICATIONS TO SUPERCRITICAL GALTON-WATSON PROCESSES

A. JOFFE

Centre de Recherches Mathématiques
Université de Montréal
Montreal, Quebec

1. INTRODUCTION

The study of the family structure of branching processes presents intrinsic interest ([1]) and is also intimately related to the study of branching random walks (cf. [2] in the critical case and [4] in the supercritical case). In this paper, motivated by properties of the supercritical Galton-Watson processes, we study first deterministic trees in the spirit of [4].

2. DEFINITIONS AND NOTATION

In order to facilitate the reader's task, we recall some of the notations and definitions of [4].

For the purpose of this paper, a tree $(T,\leq)$ is a set T with a partial order $\leq$ satisfying the following conditions:

(i) Each nonempty subset of T has an infimum in T. Let $0 = \inf T$.

(ii) For each $\alpha \in T$, the set $P_\alpha = \{\beta | 0 < \beta \leq \alpha\}$ is finite and totally ordered by $\leq$.

(iii) For each positive integer n the set $T_n = \{\alpha | |P_\alpha| = n\}$ is finite and $|T_n| \geq 1$. (For any set A, $|A|$ denotes the cardinality of A.) Let $\xi_n = |T_n|$.

Elements of T_n will be referred to as paths of length n or members of the nth generation. For $\alpha \in T_n$, we set $|\alpha| = n$. Define $\alpha(n,k)$ by

$$\alpha(n,k) = |\{(\tau,\tau') \in T_n \times T_n ; \tau \wedge \tau' \in T_k\}|, \tag{2.1}$$

where as usual $\wedge$ denotes the inf. Clearly we have

$$\sum_{k=0}^{n} \alpha(n,k) = \xi_n^2. \tag{2.2}$$

For any $\tau \in T$, we define the tree $(T^\tau, \leq)$ where

$$T^\tau = \{\alpha | \tau \leq \alpha\}, \tag{2.3}$$

clearly T^τ inherits the structure of a tree where τ plays the role of 0. We denote by $\alpha_\tau(n,k)$ the quantities corresponding (2.1) for T^τ, similarly let

$$T_n^\tau = T^\tau \cap T_n \quad \text{and} \quad \xi_\tau(n) = |T_n^\tau|. \tag{2.4}$$

Let $p_{n,k} = \alpha(n,k)/\xi^2(n)$. We say that the tree T is regular if:

$$\lim_{n\uparrow\infty} p_{n,k} = \pi_k \quad \text{with} \quad \sum_{k=0}^{\infty} \pi_k = 1. \tag{2.5}$$

For any integer $m \geq 2$, we define the m-adic tree to be the one for which all $\xi_\tau(k+1) = m$ where k denotes the length of τ. For the m-adic tree we have

$$\alpha(n,k) = (m-1)m^{2n-k-1} \qquad \text{if} \quad 0 \leq k < n, \tag{2.6.1}$$

$$\alpha(n,n) = m^n, \tag{2.6.2}$$

3. A THEOREM

Motivated by the almost sure properties of supercritical Galton-Watson processes, we introduce the following definition:

For any real number $m \geq 1$ a weakly m-adic tree is a tree T for which there exists a sequence of strictly positive numbers $\{c_n\}$ such that

(i) $\lim_{n\uparrow\infty} (c_{n+1}/c_n) = m$

(ii) for any $\tau \in T$ the sequence $\xi_n(\tau)/c_n$ has a limit as $n \uparrow \infty$ which is denoted by $W(\tau)$ with $0 < W(\tau) < \infty$.

Observe that the $W(\tau)$ satisfy the following relations:

$$W(\tau) = \frac{1}{m^k} \sum_{\xi \in T_k^\tau} W(\xi). \tag{3.1}$$

<u>Theorem.</u>

Let T be a weakly m-adic tree, the quantities $p(n,k)$ have a limit π_k as n goes to infinity. The π_k may be identified as:

$$\pi_k = a_k - a_{k-1} \tag{3.2}$$

where

$$a_k = \frac{\sum_{\tau \in T_k} W^2(\tau)}{m^{2k} W^2(0)} . \tag{3.3}$$

<u>Proof.</u>

A simple counting argument gives:

$$\alpha(n,k) = \sum_{\tau \in T_k} \sum_{\substack{\sigma,\sigma' \in T_1^\tau \\ \sigma \neq \sigma'}} \xi_{n-k-1}(\sigma)\, \xi_{n-k-1}(\sigma'). \tag{3.4}$$

Multiplying both sides of this equation by $[(c_n/c_{n-k-1})\xi_n]^2$, taking the limit as $n\uparrow\infty$ and using the properties of the tree we obtain:

$$\pi_k = \lim_{n\uparrow\infty} \alpha(n,k)/\xi_n^2 = \sum_{\tau\in T_k} \sum_{\substack{\sigma,\sigma'\in T_1^\tau \\ \sigma\neq\sigma'}} \frac{W(\sigma)\ W(\sigma')}{m^{2k+2}\ W^2(\sigma)} . \qquad (3.5)$$

From

$$\sum_{\substack{\sigma,\sigma'\in T_1^\tau \\ \sigma\neq\sigma'}} W(\sigma)\ W(\sigma') = \sum_{\sigma,\sigma'\in T_1^\tau} W(\sigma)\ W(\sigma') - \sum_{\sigma\in T_1^\tau} W^2(\sigma)$$

$$= \left[\sum_{T_1^\tau} W(\sigma)\right]^2 - \sum_{T_1^\tau} W^2(\sigma)$$

$$= [mW(\tau)]^2 - \sum_{\sigma\in T_1^\tau} W^2(\sigma) \qquad (3.6)$$

the conclusion follows.

4. MEASURES ON TREES

To each tree T we associate the space X of all its maximal chains (i.e. its infinite branches). For each $\tau \in T$ let $E_\tau \subset X$, the subset of those chains which contain τ; $\{E_\tau\}_{\tau\in T_k}$ is obviously a partition of X. As k increases these partitions become finer. Therefore the E_τ may be taken as a base for a topology on X; it can be shown that X is a compact metrizable totally disconnected space for this topology.

The formula $\nu(E_\tau) = \Pi_{\alpha<\tau}(1/\xi_1(\alpha))$ defines an additive positive set function on the sets making up the partitions which extend to a Borel probability measure.

For a weakly m-adic tree there is another natural measure defined by

$$\mu(E_\tau) = \frac{W(\tau)}{m^{|\tau|}} . \qquad (4.1)$$

In fact (4.1) establishes a 1-1 correspondance between additive set functions on $\{E_\tau\}$ and functions satisfying (3.1).

In terms of the measure μ the a_k identified in (3.3) are given by

$$a_k = \sum_{\tau \in T_k} \frac{\mu^2(E_\tau)}{\mu^2(X)}, \tag{4.2}$$

i.e. the quadratic variation of μ on the kth partition. Thus $a_k \downarrow 0$ if and only if the quadratic variation of μ is zero, i.e. if and only if μ is non-atomic. We conjecture that this is the case (G.L. O'Brien, York University, has recently produced a nice counterexample to that conjecture).

It is well known that supercritical Galton-Watson trees conditioned on non-extinction are almost surely weakly m-adic. From Theorem 5 of [4] and Theorem 1 of [5] it follows that our conjecture is true for those Galton-Watson trees with $EZ\log Z < \infty$.

It would be of interest to know under which conditions μ is absolutely continuous with respect to ν.

REFERENCES

1. W.Buhler, Proc.Sixth Berkeley Symp., III:463 (1972).
2. R.Durett, The Genealogy of Critical Branching Processes (to appear).
3. A.Joffe, L. Le Cam, J.Neveu, C.R.Acad.Sc., Paris, Série A 277: 963 (1973).
4. A.Joffe, A.Moncayo, Advances in Math. 10:401 (1973).
5. N. Kaplan, S.Asmussen, J.Stoch.Processes Appl. 4:15 (1976).

DIFFUSION APPROXIMATIONS FOR BRANCHING PROCESSES

Thomas G. Kurtz

Department of Mathematics
University of Wisconsin-Madison
Madison, Wisconsin

0. INTRODUCTION

The purpose of this paper is to illustrate how semigroup approximation theorems can be used to obtain diffusion and related approximations for branching processes. The diffusion approximations we will give are extensions in two directions of those first formulated by Feller [5] and made rigorous by Jirina [10] (see also Lindvall [14]). The first will be approximations of Feller type to branching processes in random environments (BPRE) and the second will be a diffusion approximation for a continuous time, two dimensional, Markov branching process. The BPRE approximation was conjectured by Keiding [11] and made rigorous in the case of independent environments for each generation by Helland [7].

Other extensions of these approximations include work of Jagers [9] in the case of age dependence, and of Grimvall [6] for general offspring distributions (e.g. in the domain of attraction of

* Work supported in part by the National Science Foundation.

a stable law). See also Helland [8] for a different approach to Grimvall's results, and Lamperti [13] for related results.

In Section I we state the general approximation results we will need. These are modifications of results in Kurtz [12] and will be discussed in detail elsewhere. In Section II we consider BPRE and in Section III we consider the two dimensional Markov branching process.

I. GENERAL APPROXIMATION THEOREMS

The approximation techniques we will describe are based on the semigroup associated with a Markov process. If $X(t)$ is a temporally homogeneous Markov process with values in a locally compact, separable metric space E having transition function $P(t, x\Gamma)$, then

$$T(t)f(x) \equiv E_x(f(X(t))) = \int f(y)P(t,x,dy) \tag{1.1}$$

defines a semigroup of operators on $B(E, \mathcal{B})$, the space of bounded Borel measurable functions on E. We take the sup norm on $B(E, \mathcal{B})$. Then $T(t)$ satisfies

$$T(t+s) = T(t)T(s) \tag{1.2}$$

and

$$\|T(t)\| \leq 1 \ . \tag{1.3}$$

A semigroup satisfying (1.3) is called a contraction semigroup.

$T(t)$ is strongly continuous on a subspace $K \subset B(E, \mathcal{B})$ if

$$T(t) : K \to K \qquad \text{all} \qquad t \geq 0 \tag{1.4}$$

and

(1.5) $$\lim_{t \to 0} T(t)f = f \quad \text{(strong limit) all} \quad f \in K .$$

The object of primary interest in the study of a strongly continuous semigroup is the infinitesimal operator defined by

(1.6) $$Af = \lim_{t \to 0} \frac{T(t)f - f}{t}$$

whenever the limit exists. Necessary and sufficient conditions for an operator A to be the infinitesimal generator of a strongly continuous semigroup are given by the Hille-Yosida Theorem (see Dynkin [4] page 30). A subset D of the domain of A, $\mathcal{D}(A)$ is called a core for A if the closure of $A|_D$ (A restricted to D) is A.

Lemma (1.7) Let T(t) be a strongly continuous contraction semigroup on a Banach space K. Then $D \subset \mathcal{D}(A)$ is a core for the infinitesimal operator A if and only if D and $\mathcal{R}(\lambda - A|_D)$ are dense in K for some $\lambda > 0$.

If $D \subset \mathcal{D}(A)$ is dense in K then a sufficient condition for D to be a core is that $T(t) : D \to D$.

We will be interested in the semigroup of a Markov process restricted to a subspace on which it is strongly continuous.

The following lemma gives two examples of cores which we will need in our applications.

Lemma (1.8) (a) $C_0^\infty[0, \infty)$, the space of infinitely differentiable functions with compact support in $[0, \infty)$, is a core for the generator $Af(x) = \alpha x f''(x)$ in $\hat{C}([0, \infty)$, the space of continuous functions vanishing at ∞.

(b) $C_0^\infty([0, \infty) \times (-\infty, \infty))$ is a core for the generator $Af(x, y) = x(af_{xx} + 2bf_{xy} + cf_{yy})$ in $\hat{C}([0, \infty) \times (-\infty, \infty))$.

Proof: We prove part (b). The proof of (a) is similar but much simpler. We begin by observing that the differential equation $u_t = Au$ has solutions of the form $u(t, x, y) = \exp\{-\alpha(t)x + i\beta y\}$ where the real part of $\alpha(t)$ is positive and β is real. In particular

(1.9) $$\dot{\alpha}(t) = -a\alpha^2(t) + ib\alpha(t)\beta + c\beta^2,$$

and we may select $\alpha(0)$ to be any complex number with positive real part and β to be any real number. Let D_K be the space of linear combinations of functions of the form $\exp\{-\alpha x + i\beta y\}$ with $\operatorname{Re}\alpha > 0$ and β real, and let K be the closure of D_K in the sup norm. Since $T(t) : D_K \to D_K$, D_K is a core for A in the Banach space K. In particular $\mathcal{R}(\lambda - A|_{D_K})$ is dense in K. Let $D = C_0^\infty([0,\infty) \times (-\infty,\infty))$, and suppose that $\mathcal{R}(\lambda - A|_D)$ is not dense in $\hat{C}$. Since the bounded linear functionals on $\hat{C}$ are the finite signed measures, there must be a nonzero finite signed measure μ such that

$$\int (\lambda f - Af)d\mu = 0$$

for every $f \in D$. Let $\mathfrak{J}$ be nonnegative and infinitely differentiable on $[0,\infty) \times (-\infty,\infty)$ such that $\mathfrak{J}(x,y) = 1$ for $x^2 + y^2 \le 1$ and $\mathfrak{J}(x,y) = 0$ for $x^2 + y^2 \ge 2$. Let $g \in D_K$ and let $g_n(x,y) = \mathfrak{J}(x/n, y/n)g(x,y)$. Then $\sup_n \|g_n\| < \infty$, $\sup_n \|Ag_n\| < \infty$ $\lim_{n\to\infty} g_n(x,y) = g(x,y)$, $\lim_n Ag_n(x,y) = Ag(x,y)$. Since $g_n \in D$, $\int (\lambda g_n - Ag_n)d\mu = 0$ and hence $\int (\lambda g - Ag)d\mu = 0$ for every $g \in D_K$. But $\mathcal{R}(\lambda - A|_{D_K})$ is dense in K so $\int f d\mu = 0$ for every $f \in K$. Since K contains the functions $\exp\{-\alpha x + i\beta y\}$, this implies $\mu = 0$ which is a contradiction.

Let $(\Omega, \mathfrak{F}, P)$ be a probability space, $\{\mathfrak{F}_t\}$ an increasing family of sub-σ-algebras, and $X(t)$, a stochastic process on $(\Omega, \mathfrak{F}, P)$ that is Markovian with respect to $\{\mathfrak{F}_t\}$.

The infinitesimal operator describes the behavior of the Markov process over small time intervals. Specifically

(1.10) $$\lim_{\varepsilon\to 0} \frac{1}{\varepsilon} E(f(X(t+\varepsilon)) - f(X(t)) | \mathfrak{F}_t) = Af(X(t)) \text{ a.s.}$$

We are interested in comparing a sequence of processes $\{X_n(t)\}$ which are not necessarily Markovian to the Markov process $X(t)$. In order to do this we want to consider how close the "infinitesimal behavior" of $X_n(t)$ is to that of $X(t)$ which is described by A. If the behavior is similar, then for small ε

$$\frac{1}{\varepsilon}E(f(X_n(t+\varepsilon)) - f(X_n(t))|\mathfrak{F}_t) \tag{1.11}$$

should be approximately $Af(X_n(t))$. Alternatively, we could take a limit in (1.11) (if it exists) as in (1.10) and compare the limit to $Af(X_n(t))$. With this in mind we have the following theorem:

<u>Theorem</u> (1.12). Let E be a locally compact separable metric space and let $X(t)$ be a Markov process with values in E whose semigroup $T(t)$ is strongly continuous on $\hat{C}(E)$. Let D be a core for the infinitesimal operator A of $T(s)$ in $\hat{C}(E)$. Let $\{X_n(t)\}$ be a sequence of right continuous E valued processes adapted to an increasing family of σ-algebras $\{\mathfrak{F}_t\}$. Suppose for every $f \in D$ and every $T > 0$ one of the following holds:

(a) there exist continuous functions f_n (not necessarily bounded) and $\varepsilon_n > 0$ satisfying

$$\sup_n E\left(\sup_{t \le T} |f_n(X_n(t))|\right) < \infty \tag{1.13}$$

$$\sup_n \sup_{t \le T} E\left(\left|\varepsilon_n^{-1} E(f_n(X_n(t+\varepsilon_n)) - f_n(X_n(t))|\mathfrak{F}_t)\right|\right) < \infty, \tag{1.14}$$

$$\lim_{n \to \infty} E\left(\left|\varepsilon_n^{-1} \int_0^{\varepsilon_n} E(f_n(X_n(t+s))|\mathfrak{F}_t)ds - f(X_n(t))\right|\right) = 0, \tag{1.15}$$

and

$$\lim_{n \to \infty} E\left(\left|\varepsilon_n^{-1} E(f_n(X_n(t+\varepsilon_n)) - f_n(X_n(t))|\mathfrak{F}_t) - Af(X_n(t))\right|\right) = 0; \tag{1.16}$$

(b) there exist right continuous processes $f_n(t)$ and $h_n(t)$ satisfying

$$\sup_n \sup_{t \le T} E(|f_n(t)|) < \infty, \tag{1.17}$$

$$(1.18)\qquad \sup_n \sup_{0<s,\ 0\le t\le T} E(|s^{-1}E(f_n(t+s) - f_n(t)|\mathfrak{F}_t)|) < \infty ,$$

$$(1.19)\qquad \lim_{s\to 0} E(|s^{-1}E(f_n(t+s) - f_n(t)|\mathfrak{F}_t) - h_n(t)|) = 0 \qquad \text{for all } t \le T ,$$

$$(1.20)\qquad \lim_{n\to\infty} E(|f_n(t) - f(X_n(t))|) = 0 \qquad \text{for all } t \le T ,$$

$$(1.21)\qquad \lim_{n\to\infty} E(|h_n(t) - Af(X_n(t))|) = 0 \qquad \text{for all } t \le T .$$

If $X_n(0)$ converges in distribution to $X(0)$, then the finite dimensional distributions of X_n converge to those of X.

Remark: If $X_n(t)$ is a Markov process with infinitesimal operator A_n and $f_n(t) = g_n(X_n(t))$ where $g_n \in \mathcal{D}(A_n)$ then $h_n(t) = A_n g_n(X_n(t))$.

Application of the above theorem requires information about the distribution of $X_n(t+\varepsilon_n)$ conditioned on $\mathfrak{F}_t$. When this type of information is available the following lemma can be useful in verifying tightness in $D_E[0,\infty)$. (See Billingsley [2] and Lindvall [15].) A slightly more general version of the lemma is in Kurtz [12] and a closely related result is in Billingsley [3]. $\rho(x,y)$ denotes the metric on E and $r(x,y) = 1 \wedge \rho(x,y)$.

Lemma (1.22). Let $\{X_n(t)\}$ be a sequence of processes with sample paths in $D_E[0,\infty)$. Let $\beta > 0$ and suppose

(A) for each $\eta > 0$ and $T > 0$ there is a compact set $K \subset E$ such that $\varliminf_{n\to\infty} P\{X_n(t) \in K \text{ all } t \le T\} > 1 - \eta$.

(B) for each $T > 0$, $\delta > 0$ and n there are random variables $\gamma_n(\delta)$ such that

$$E(r^{\beta}(X_n(t+u), X_n(t))|\mathfrak{F}_t) \le E(\gamma_n(\delta)|\mathfrak{F}_t) \quad \text{a.s.}$$

for all $0 \le t \le T$ and $0 \le u \le \delta$. If $\lim_{\delta \to 0} \overline{\lim_{n \to \infty}} E(\gamma_n(\delta)) = 0$ then the sequence $\{X_n(t)\}$ is tight. (i.e. a subsequence converges weakly in the Skorohod topology on $D[0,\infty)$.)

The following gives conditions for tightness even more closely related to the conditions of Theorem (1.12).

<u>Lemma</u> (1.23). Let $\{X_n(t)\}$ be a sequence of processes with sample paths in $D_E[0,\infty)$ and let D be the subset of $f \in \hat{C}(E)$ such that for every $\varepsilon > 0$ and $T > 0$ there are right continuous processes f_n and h_n satisfying the following:

(1.24) $$\overline{\lim_{n \to \infty}} E(\sup_{t \le T} |f_n(t) - f(X_n(t))|) < \varepsilon ,$$

(1.25) $$\sup_{0 < s,\ 0 \le t \le T} E(|s^{-1} E(f_n(t+s) - f_n(t) | \mathcal{F}_t)|) < \infty ,$$

(1.26) $$\lim_{s \to 0} E(|s^{-1} E(f_n(t+s) - f_n(t) | \mathcal{F}_t) - h_n(t)|) = 0 ,$$

and either

(1.27) $$\overline{\lim_{n \to \infty}} \sup_{t \le T} E(|h_n(t)|^p) < \infty$$

for some $p > 1$ or

(1.28) $$\overline{\lim_{n \to \infty}} E(\sup_{t \le T} |h_n(t)|) < \infty .$$

(a) If Condition (A) of Lemma (1.22) is satisfied and D is dense in $\hat{C}(E)$ then $\{X_n(t)\}$ is tight.

(b) If the finite dimensional distributions of $X_n(t)$ converge to those of a process $X(t)$ with sample paths in $D_E[0,\infty)$ and D is dense in $\hat{C}(E)$ then $\{X_n(t)\}$ converges weakly to $X(t)$.

<u>Remark</u>: Note that in (b), Condition (A) does not have to be verified.

II. BRANCHING PROCESSES WITH RANDOM ENVIRONMENTS

Let $Z_n(k)$ be a sequence of stochastic processes defined on a probability space $(\Omega, \mathfrak{F}, P)$ and satisfying the interative formula

$$(2.1) \qquad Z_n(k) = \sum_{i=1}^{Z_n(k-1)} \xi_{i,k}^{(n)} \qquad k \geq 1$$

where the $\xi_{ik}^{(n)}$ are nonnegative integer valued random variables. We assume that there is a σ-algebra $\mathcal{E} \subset \mathfrak{F}$, intuitively containing all information about the environment, such that, conditioned on $\mathcal{E}$, the $\xi_{ik}^{(n)}$ are independent and for fixed k and n they are identically distributed. In addition we assume $Z_n(0)$ is independent of $\mathcal{E}$. The processes are called branching processes with random environments. (See for example Athreya and Ney [1] and Tanny [16].)

We define the following random variables

$$(2.2) \qquad m_k^{(n)} = E(\xi_{ik}^{(n)} | \mathcal{E})$$

$$\alpha_k^{(n)} = E((\xi_{ik}^{(n)}/m_k^{(n)} - 1)^2 | \mathcal{E})$$

$$\gamma_k^{(n)} = E(|\xi_{ik}^{(n)}/m_k^{(n)} - 1|^3 | \mathcal{E}),$$

and the processes

$$(2.3) \qquad X_n(t) = Z_n([nt])/n$$

$$(2.4) \qquad M_n(t) = \sum_{k=1}^{[nt]} \log m_k^{(n)}$$

$$(2.5) \qquad Y_n(t) = X_n(t) e^{-M_n(t)}$$

$$(2.6) \qquad A_n(t) = \frac{1}{n} \sum_{k=1}^{[nt]} \alpha_k^{(n)}$$

and

$$(2.7) \qquad G_n(t) = \frac{1}{n^{3/2}} \sum_{k=1}^{[nt]} Y_k^{(n)} \quad .$$

We define $\tau_n(t)$ as the solution of

$$(2.8) \qquad t = \int_0^{\tau_n(t)} \alpha_{[ns]+1} e^{-M_n(s)} ds$$

and

$$(2.9) \qquad W_n(t) = Y_n(\tau_n(t)) \ ,$$

for all $t < \int_0^{\infty} \alpha_{[ns]+1} e^{-M_n(s)} ds$.

While we are primarily interested in the behavior of $X_n(t)$, we will first prove weak convergence of $W_n(t)$ and from this conclude convergence of $X_n(t)$. The reason for this is that $W_n(t)$ is almost independent of the environment. We can therefore prove convergence of $W_n(t)$ under very general assumptions about the environment.

We begin by treating the case of nonrandom, but time dependent environments i.e. the $m_k^{(n)}$ etc. are constants and $M_n(t)$, $A_n(t)$ and $G_n(t)$ are nonrandom functions. From this theorem we are able to conclude almost immediately the corresponding results in the random case.

Theorem (2.10). Suppose $M_n(s)$ converges in the Skorohod Topology to a function $M(s)$ in $D[0,\infty)$, that $\lim_{n\to\infty} A_n(t) = A(t)$ exists, where $A(t)$ is absolutely continuous and strictly increasing, and that $\lim_{n\to\infty} G_n(t) = 0$ for every t. Define $\tau(t)$ as the solution of

$$(2.11) \qquad t = \int_0^{\tau(t)} e^{-M(s)} A'(s) ds$$

for $t < \int_0^\infty e^{-M(s)} A'(s)ds \equiv T_0$.

Let $W(t)$ be a diffusion with generator $\frac{1}{2} wf''(w)$. If $W_n(0) \equiv X_n(0)$ converges in distribution to $W(0)$, then $W_n(t)$ converges weakly in $D[0,T]$ to $W(t)$ for every $T < T_0$ and $X_n(t)$ converges weakly in $D[0,\infty)$ to

$$X(t) \equiv W(\tau^{-1}(t))e^{M(t)} \quad . \tag{2.12}$$

Before proving Theorem (2.10) we show how to obtain the corresponding conclusion in the case of random environments.

<u>Theorem</u> (2.13) Let $X_n(t)$ be a sequence of BPRE normalized as in (2.3). Let $M_n(t)$, $A_n(t)$ and $G_n(t)$ be given by (2.4), (2.6) and (2.7), and $W_n(t)$ by (2.9). Suppose that $(M_n(t), A_n(t))$ converges weakly in $D_{\mathbb{R}^2}[0,\infty)$ to a process $(M(t), A(t))$ and that $G_n(t) \to 0$ in probability for every t . Suppose that $A(t)$ is absolutely continuous a.s. Define $\tau(t)$ as the solution of

$$t = \int_0^{\tau(t)} e^{-M(s)} A'(s)ds \quad . \tag{2.14}$$

Let $W(t)$ be a diffusion with generator $\frac{1}{2} wf''(w)$ that is independent of $(M(t), A(t))$. Suppose $W_n(0) \equiv X_n(0)$ is independent of the environment and $W_n(0)$ converges in distribution to $W(0)$.

(a) If $\int_0^\infty e^{-M(s)} A'(s)ds = \infty$ a.s. then

$(W_n(t), M_n(\tau_n(t)), A_n(\tau_n(t)))$ converges weakly to $(W(t), M(\tau(t)), A(\tau(t)))$.

(b) If $P\{\int_0^\infty e^{-M(s)} A'(s)ds < \infty\} > 0$, define

(2.15) $$\theta_N^{(n)} = \int_0^N e^{-M_n(s)} \alpha_{[ns]}^{(n)} ds$$

and

$$\theta_N = \int_0^N e^{-M(s)} A'(s)ds ,$$

then $(W_n(t \wedge \theta_N^{(n)}), M_n(\tau_n(t \wedge \theta_N^{(n)})), A_n(\tau_n(t \wedge \theta_N^{(n)})))$ converges weakly to $(W(t \wedge \theta_N), M(\tau(t \wedge \theta_N)), A(\tau(t \wedge \theta_N)))$ for every $N > 0$. In either case $X_n(t)$ converges weakly to $W(\tau^{-1}(t))e^{M(t)}$.

Proof: Since we are only interested in convergence in distribution we can, without loss of generality, assume $(M_n(t), A_n(t))$ converges a.s. in the Skorohod topology on $D_{\mathbb{R}^2}[0,\infty)$ to $(M(t), A(t))$ and that $\lim_{n \to \infty} G_n(t) = 0$ a.s. for every t. This implies $\lim_{n \to \infty} \theta_N^{(n)} = \theta_N$ as well. Under the conditions of (b), we can modify the environment of $s > N$ so that

$$\int_0^\infty e^{-M_n(s)} \alpha_{[ns]+1} ds = \int_0^\infty e^{-M(s)} A'(s)ds = \infty \quad \text{a.s.}$$

Since this modification does not affect $W_n(t \wedge \theta_N^{(n)})$, $M_n(\tau_n(t \wedge \theta_N^{(n)}))$ and $A_n(\tau_n(t \wedge \theta_N^{(n)}))$, (b) follows from (a).

Let $\mathcal{E}$ be the σ-algebra generated by all the environments. From a careful construction of the BPRE and Theorem (2.10) one can conclude that for every bounded continuous functional F on $D_{\mathbb{R}}[0,\infty)$

(2.16) $$\lim_{n \to \infty} E(F(W_n) | \mathcal{E}) = E(F(W)) \quad \text{a.s.}$$

Consequently, it follows that

(2.17) $$\lim_{n\to\infty} E(F(W_n)H(M_n, A_n)) = E(F(W))E(H(M, A))$$

for all bounded continuous functionals F on $D_{\mathbb{R}}[0,\infty)$ and H on $D_{\mathbb{R}^2}[0,\infty)$. The weak convergence and the independence of W and (M, A) follow.

Corollary (2.18). If $(M(t), A'(t))$ is a Markov process with generator $Bf(y, z)$, then $(W(\tau^{-1}(t)), M(t), A'(t))$ is a Markov process with generator of the form

(2.19) $$Af(w, y, z) = \frac{1}{2} ze^{-y} wf_{ww}(w, y, z) + Bf(w, y, z) .$$

If $A'(t) \equiv \alpha$ and $M(t)$ is a Brownian motion with generator $\frac{1}{2} af'' + bf'$, then $X_n(t)$ is a diffusion with generator

(2.20) $$= Af(x) = \frac{1}{2} (\alpha x + ax^2)f_{xx} + (\frac{1}{2} a + b)xf_x \quad .$$

Remark: This last case, $A'(t) = \alpha$ and $M(t)$ a Brownian motion, is the situation that arises in the result conjectured by Keiding and proved by Helland for independent environments. Helland obtains his result without assuming that the offspring distributions have finite third moments. Sufficient conditions for the diffusion limit (2.20) in the independent environments case under our hypotheses would be as follows:

$$\lim_{n\to\infty} n\, E(\log m_k^{(n)}) = b$$

$$\lim_{n\to\infty} n\, E((\log m_k^{(n)})^2) = a$$

$$\lim_{n\to\infty} n\, P\{|\log m_k^{(n)}| > \varepsilon\} = 0 \quad \text{for every } \varepsilon > 0$$

$$\lim_{n\to\infty} E(\alpha_k^{(n)}) = \alpha$$

$$\lim_{n\to\infty} n^{-1/2} E(\gamma_k^{(n)}) = 0$$

The first three conditions imply the weak convergence of $M_n(t)$ to the Brownian motion with generator $\frac{1}{2}af'' + bf'$ and the last two imply $A_n(t) \to \alpha t$ and $G_n(t) \to 0$ in probability. Helland's result (without third moments) can be obtained from ours by a truncation argument (i.e. replace $\xi_{ik}^{(n)}$ by $[\xi_{ik}^{(n)} \wedge \sqrt{n}\,]$).

We need the following lemmas for the estimates in the proof of Theorem (2.10). Suppressing the obvious dependence on n we define $t^* = [nt]/n$.

Lemma (2.21). The following processes are martingales with respect to $\mathfrak{F}_t \equiv \sigma(X_n(s), s \le t) \vee \mathcal{E}$.

(a) $Y_n(t) \equiv X_n(t) e^{-M_n(t)}$

(b) $Y_n^2(t) - \int_0^{t^*} Y_n(s)\alpha_{[ns]+1}^{(n)} e^{-M_n(s)}\, ds$.

Proof: The fact that $Y_n(t)$ is a martingale is standard. To see that (b) is a martingale observe that

$$E((Y_n(\tfrac{k+1}{n}) - Y_n^2(\tfrac{k}{n})) \mid \mathfrak{F}_{k/n}) = Y_n(\tfrac{k}{n})\, e^{-M_n(k/n)} \alpha_{k+1}^{(n)}/n \quad (2.22)$$

Lemma (2.23). Let τ be a stopping time with respect to $\{\mathfrak{F}_t\}$. Then

$$\begin{aligned} &|Y_n(t+\tau) - Y_n(\tau)|^3 \\ &- \int_{\tau^*}^{(\tau+t)^*} \frac{3}{\sqrt{n}}\, Y_n^{3/2}(s)\gamma_{[ns]+1}^{(n)} e^{-\frac{3}{2}M_n(s)}\, ds \\ &- \int_{\tau^*}^{(\tau+t)^*} 3Y_n(s)\alpha_{[ns]+1}^{(n)} \,|\, Y_n(s) - Y_n(\tau) \,|\, e^{-M_n(s)}\, ds \end{aligned} \quad (2.24)$$

is a supermartingale.

Proof: If $U_0 = 0$, $U_1, U_2 \cdots$ is a martingale with $E(|U_k|^3) < \infty$, then

$$(2.25)\qquad |U_m|^3 - \sum_{k=0}^{m-1} E(|U_{k+1}|^3 - |U_k|^3 \,|\mathfrak{F}_k)$$

is a martingale and since

$$(2.26)\qquad E(|U_{k+1}|^3 - |U_k|^3 \,|\mathfrak{F}_k)$$

$$\le E(|U_{k+1} - U_k|^3 \,|\mathfrak{F}_k)$$

$$+ 3\,E(|U_{k+1} - U_k|^2 \,|U_k|\, |\mathfrak{F}_k)$$

it follows that

$$(2.27)\qquad |U_m|^3 - \sum_{k=0}^{m-1} E(|U_{k+1} - U_k|^3 \,|\mathfrak{F}_k)$$

$$- 3 \sum_{k=0}^{m-1} E(|U_{k+1} - U_k|^2 \,|U_k|\, |\mathfrak{F}_k)$$

is a supermartingale.

If the increments $\xi_k = U_k - U_{k-1}$ are independent and identically distributed, we have

$$(2.28)\qquad E(|U_m|^3) \le mE(|\xi_1|^3) + 3E(\xi_1^2) \sum_{k=0}^{m-1} E(|U_k|)$$

$$\le mE(|\xi_1|^3) + 3E(\xi_1^2)^{3/2} \sum_{k=0}^{m-1} k^{1/2}$$

$$\le 3m^{3/2}\, E(|\xi_1|^3) \ .$$

Inequality (2.28) implies that for any stoping time τ

$$(2.29)\qquad E(|Y_n(\tau+\tfrac{1}{n}) - Y_n(\tau)|^3\,|\mathcal{F}_\tau)$$
$$\le \frac{3}{n^3} e^{-3M_n(\tau)} \gamma^{(n)}_{[n\tau]+1} Z_n^{3/2}([n\tau])$$
$$= \frac{3}{n^{3/2}} e^{-\frac{3}{2}M_n(\tau)} \gamma^{(n)}_{[n\tau]+1} Y_n^{3/2}(n\tau)\ .$$

We now apply (2.27) with $U_m = Y_n(\tau + \frac{m}{n}) - Y_n(\tau)$ to obtain (2.24), using (2.29) to estimate the first summation in (2.27) by the first integral in (2.24). The second summation in (2.27) will equal the second integral in (2.24), since

$$E(|Y_n(\tau+\tfrac{1}{n}) - Y_n(\tau)|^2\,|\mathcal{F}_\tau) = Y_n(\tau)\alpha_{[n\tau]+1} e^{-M_n(\tau)}\ .$$

<u>Lemma</u> (2.30).

(a) $E(\sup_{s\le T} W_n^2(s)) \le 2[E(W_n^2(0)) + TE(W_n(0))]$

(b) $E(\sup_{t\le s\le t+\varepsilon} (W_n(s) - W_n(t))^2\,|\mathcal{F}_t)$
$$\le 2W_n(t)[\varepsilon + \frac{1}{n}\alpha_{[n\tau_n(t)]+1} e^{-M_n(\tau_n(t))}]$$

<u>Proof</u>: Since $W_n(t)$ is a martingale these inequalities follow from Doob's inequality and (2.21) (b).

<u>Proof</u> of Theorem (2.10): By modifying the environment for large times we may assume that $\int_0^\infty \alpha_{[ns]+1} e^{-M_n(s)} ds = \int_0^\infty e^{-M(s)} A'(s)ds = \infty$, and by an appropriate truncation argument we may assume $\sup_n E(W_n(0)^2) < \infty$.

Define $\mathcal{G}_t^{(n)} = \mathcal{F}_{\tau_n(t)}$, let $f \in C_0^\infty[0,\infty)$ and fix $T > 0$. Then, since W_n is a martingale,

$$(2.31)\quad E(f(W_n(t+\varepsilon)) - f(W_n(t))|\mathcal{G}_t^{(n)})$$

$$= \frac{1}{2}E((W_n(t+\varepsilon) - W_n(t))^2|\mathcal{G}_t^{(n)})f''(W_n(t))$$

$$+ 0(E(|W_n(t+\varepsilon) - W_n(t)|^3|\mathcal{G}_t^{(n)}))$$

$$= \frac{1}{2}E(\int_{\tau_n^*(t)}^{\tau_n^*(t+\varepsilon)} Y_n(s)\alpha_{[ns]+1}e^{-M_n(s)}ds|\mathcal{G}_t^{(n)})f''(W_n(t))$$

$$+ 0(E(\int_{\tau_n^*(t)}^{\tau_n^*(t+\varepsilon)} \frac{3}{\sqrt{n}}Y_n^{3/2}(s)Y_{[ns]+1}^{(n)}e^{-3/2\,M_n(s)}ds|\mathcal{G}_t^{(n)}))$$

$$+ 0(E(\int_{\tau_n^*(t)}^{\tau_n^*(t+\varepsilon)} 3Y_n(s)\alpha_{[ns]+1}^{(n)}|Y_n(s) - Y_n(\tau_n(t))|e^{-M_n(s)}ds|\mathcal{G}_t^{(n)}))\ .$$

For $t \le T$, the integral in the first error term is bounded by

$$(2.32)\qquad 3\sup_{s\le T} W_n^{3/2}(s)\sup_{s\le\tau_n(T)} e^{-3/2\,M_n(s)}G_n(\tau_n(T))\ .$$

Since $Y_n(s) = Y_n(\tau_n(t))$ for $\tau_n^*(t) \le s \le \tau_n(t)$, the lower limit of the integral in the second error term may be replaced by $\tau_n(t)$. This integral is then bounded by

$$(2.33)\qquad 3\varepsilon\sup_{t\le T}\sup_{t\le s\le t+\varepsilon} W_n(s)|W_n(s) - W_n(t)|\ .$$

Finally, the coefficient of f'' equals

$$(2.34)\qquad W_n(t)\int_{\tau_n^*(t)}^{\tau_n^*(t+\varepsilon)} \alpha_{[ns]+1}e^{-M_n(s)}ds$$

$$= W_n(t)(\varepsilon + 0(\frac{1}{n}\sup_{s\le\tau_n(t)}\alpha_{[ns]+1}e^{-M_n(s)}))\ .$$

From the assumptions on A_n and M_n it follows that

$$\lim_{n\to\infty} \int_0^t \alpha_{[ns]+1}\, e^{-M_n(s)} ds = \int_0^t e^{-M(s)} A'(s) ds \tag{2.35}$$

for all $t > 0$.

Since A is strictly increasing it follows that

$$\lim_n \tau_n(t) = \tau(t) \quad \text{for all} \quad t > 0 \ . \tag{2.36}$$

By the continuity of A we must have

$$\lim_{n\to\infty} \frac{1}{n} \sup_{s\le t} \alpha_{[ns]+1}\, e^{-M_n(s)} = 0 \ . \tag{2.37}$$

Since $\sup_n \tau_n(T) < \infty$, (2.37) and the assumption that $\lim_{n\to\infty} G_n(t) = 0$ imply that we may select $\varepsilon_n \to 0$ slowly enough so that

$$\lim_{n\to\infty} \varepsilon_n^{-1} \frac{1}{n} \sup_{s\le\tau_n(T)} \alpha_{[ns]+1}\, e^{-M_n(s)} = 0 \tag{2.38}$$

and

$$\lim_{n\to\infty} \varepsilon_n^{-1}\, G_n(\tau_n(T)) = 0 \ . \tag{2.39}$$

Using Lemma (2.30) to estimate (2.32) and (2.33) we conclude that

$$\lim_{n\to\infty} E(|\, \varepsilon_n^{-1}\, E(f(W_n(t+\varepsilon_n)) - f(W_n(t)) \,|\, \mathcal{G}_t^{(n)}) - \frac{1}{2} W_n(t) f''(W_n(t)) \,|\,) = 0 \ , \tag{2.40}$$

for all $t \le T$, which is condition (1.16) of Theorem (1.12) . The other conditions are easily verified. Since $C_0^\infty[0,\infty)$ is a core (Lemma (1.8)), we have convergence of the finite dimensional distributions.

Tightness follows from the inequality

$$(2.41) \qquad E((X(t+u) - X(t))^2 | \mathcal{F}_t) \le W_n(t)(u + \frac{1}{n} \sup_{s \le \tau_n(t)} \alpha_{[ns]+1} e^{-M_n(s)})$$

by taking

$$(2.42) \qquad \gamma_n(\delta) = W_n(T)(\delta + \frac{1}{n} \sup_{s \le \tau_n(T)} \alpha_{[ns]+1} e^{-M_n(s)})$$

in Lemma (1.22). Condition (A) is immediate since W_n is a nonnegative martingale.

III. DIFFUSION APPROXIMATION FOR TWO TYPE BRANCHING PROCESSES

We now consider a branching process with two types of particles, designated type 1 and type 2. Each particle lives an exponentially distributed life time with parameter λ_1 or λ_2 depending on its type. If a type 1 particle dies, it gives rise to offspring of type 1 and type 2 with the number of offspring distributed as (γ_1, γ_2). Similarly, if a type 2 particle dies, it gives rise to offspring distributed as (ψ_1, ψ_2). We will assume that $E(\gamma_i^3), E(\psi_i^3) < \infty$. Let $m_{1j} = E(\gamma_j)$ and $m_{2j} = E(\psi_j)$. We assume that the process is critical, that is the matrix

$$(3.1) \qquad M = \begin{pmatrix} m_{11} & m_{12} \\ m_{21} & m_{22} \end{pmatrix}$$

has largest eigenvalue 1, and that m_{12} and m_{21} are both nonzero. It follows that there exist vectors (ν_1, ν_2) and (μ_1, μ_2) and $\eta > 0$ satisfying

$$(3.2)\qquad \begin{pmatrix} \lambda_1(m_{11}-1) & \lambda_1 m_{12} \\ \lambda_2 m_{21} & \lambda_2(m_{22}-1) \end{pmatrix} \begin{pmatrix} \nu_1 \\ \nu_2 \end{pmatrix} = 0$$

and

$$(3.3)\qquad \begin{pmatrix} \lambda_1(m_{11}-1) & \lambda_1 m_{12} \\ \lambda_2 m_{21} & \lambda_2(m_{22}-1) \end{pmatrix} \begin{pmatrix} \mu_1 \\ \mu_2 \end{pmatrix} = -\eta \begin{pmatrix} \mu_1 \\ \mu_2 \end{pmatrix} .$$

We may assume $\nu_1, \nu_2 > 0$. (μ_1 and μ_2 will have opposite signs.)

Let (z_1, z_2) be fixed. We will consider the sequence of branching processes $(Z_1^{(n)}(t), Z_2^{(n)}(t))$ with initial population sizes $(Z_1^{(n)}(0), Z_2^{(n)}(0)) = ([nz_1], [nz_2])$. Define

$$(3.4)\qquad X_n(t) = [\nu_1 Z_1^{(n)}(nt) + \nu_2 Z_2^{(n)}(nt)]/n$$

and

$$(3.5)\qquad Y_n(t) = [\mu_1 Z_1^{(n)}(nt) + \mu_2 Z_2^{(n)}(nt)]/n \quad .$$

We note that $X_n(t)$ and $Y_n(t)e^{n\eta t}$ are martingales.

We are interested in the limiting behavior of $X_n(t)$ and $Y_n(t)$ as n goes to infinity. As might be expected from the fact that $Y_n(t)e^{n\eta t}$ is a martingale, $Y_n(t)$ degenerates an $n \to \infty$. In fact, as will be made precise, $Y_n(t)$ is essentially white noise for large n . Consequently we define a process, also a martingale,

$$(3.6)\qquad W_n(t) = Y_n(t) + \int_0^t n\eta Y_n(s)ds$$

whose limiting behavior will give us the information we want about Y_n .

Define

$$(3.7)\qquad \xi_1 = \nu_1(\gamma_1-1) + \nu_2\gamma_2 \qquad \xi_2 = \mu_1(\gamma_1-1) + \mu_2\gamma_2$$
$$\varphi_1 = \nu_1\psi_1 + \nu_2(\psi_2-1) \qquad \varphi_2 = \mu_1\psi_1 + \mu_2(\psi_2-1)$$

and

$$(3.8)\qquad \alpha^1_{ij} = E(\xi_i\xi_j)$$
$$\alpha^2_{ij} = E(\varphi_i\varphi_j) \quad .$$

Note that $E(\xi_1) = E(\varphi_1) = 0$, $E(\xi_2) = -\eta\frac{\mu_1}{\lambda_1}$ and $E(\varphi_2) = -\eta\frac{\mu_2}{\lambda_2}$.

Theorem (3.9). $(X_n(t), W_n(t))$ converges weakly to the diffusion process $(X(t), W(t))$ with $(X(0), W(0)) = (\nu_1 z_1 + \nu_2 z_2, \mu_1 z_1 + \mu_2 z_2)$ and generator

$$(3.10)\qquad Af(x,w) = \frac{x}{2}(a_{11}f_{xx}(x,w) + 2a_{12}f_{xw}(x,w) + a_{22}f_{ww}(x,w))$$

where $a_{ij} = (\lambda_1\mu_2\alpha^1_{ij} - \lambda_2\mu_1\alpha^2_{ij})/(\nu_1\mu_2 - \mu_1\nu_2)$.

Before proving the Theorem we give the following corollary which describes the behavior of $Y_n(t)$.

Corollary (3.11). For each T , $\sup_{t\le T}|Y_n(t) - Y_n(0)e^{-n\eta t}|$ converges to zero in probability. Consequently, for $0 < t_1 < t_2$, $\int_{t_1}^{t_2} n\eta Y_n(s)ds$ converges in distribution to $W(t_2) - W(t_1)$.

Remark: Since W is a random time change of a Brownian motion, i.e. $W(t) = B(\int_0^t X(s)ds)$, $n\eta Y_n(s)$ is, for large n , essentially white noise with variance depending on $X_n(s)$.

Proof of (3.11). Equation (3.6) can be inverted to give

$$(3.12)\qquad Y_n(t) - e^{-n\eta t}Y_n(0) = \int_0^t n\eta e^{-n\eta(t-s)}(W_n(t)-W_n(s))ds$$
$$+ e^{-n\eta t}(W_n(t)-W_n(0)) \ .$$

For $\varepsilon > 0$ let $w_n(\varepsilon) = \sup_{\substack{|t-s| < \varepsilon \\ t,s < T}} |W_n(t) - W_n(s)|$.

Then (3.12) implies

$$(3.13) \qquad |Y_n(t) - e^{-n\eta t} Y_n(0)| \leq \int_0^t n\eta e^{-n\eta u} w_n(u) du + e^{-n\eta t} w_n(t) \quad .$$

For $t \leq \varepsilon$ the right hand side of (3.13) is bounded by $w_n(\varepsilon)$. For $\varepsilon < t \leq T$ it is bounded by $w_n(\varepsilon) + e^{-n\eta\varepsilon} w_n(T)$. Consequently for almost every δ

$$(3.14) \qquad \overline{\lim_{n\to\infty}} \; P\{ \sup_{t \leq T} |Y_n(t) - e^{-n\eta t} Y_n(0)| > \delta\}$$

$$\leq \lim_{n\to\infty} P\{w_n(\varepsilon) > \delta\}$$

$$= P\{ \sup_{\substack{|s-t| \leq \varepsilon \\ s,t \leq T}} |W(s) - W(t)| > \delta\} \; .$$

Since ε is arbitrary and the probability on the right goes to zero as $\varepsilon \to 0$ the Corollary is proved.

Proof of Theorem (3.9). Let $f(x,w) \in C_0^\infty$. Then

$$(3.15) \quad \lim_{\varepsilon\to 0} \frac{1}{\varepsilon} E(f(X_n(t+\varepsilon), W_n(t+\varepsilon)) - f(X_n(t), W_n(t)) | \mathfrak{F}_t) \equiv$$

$$= \lambda_1 n Z_1^{(n)}(nt) E_\xi (f(X_n(t) + \tfrac{1}{n}\xi_1, W_n(t) + \tfrac{1}{n}\xi_2) - f(X_n(t), W_n(t)))$$

$$+ \lambda_2 n Z_2^{(n)}(nt) E_\varphi (f(X_n(t) + \tfrac{1}{n}\varphi_1, W_n(t) + \tfrac{1}{n}\varphi_2) - f(X_n(t), W_n(t)))$$

$$+ n\eta Y_n(t) f_w(X_n(t), W_n(t)) \; ,$$

where E_ξ and E_φ refer only to the ξ_i and φ_i . Expanding f in a Taylor series about $(X_n(t), W_n(t))$, we observe that the terms involving f_x drop out because $E(\xi_1) = E(\varphi_1) = 0$ and the terms involving f_w cancel (since $\mu_1 \frac{Z_1^{(n)}}{n} + \mu_2 \frac{Z_2^{(n)}}{n} = Y_n$) .

This leaves

$$(3.16)\qquad \frac{\lambda_1 Z_1^{(n)}(nt)}{2n}\,[\alpha_{11}^1 f_{xx}(X_n(t), W_n(t)) + 2\alpha_{12}^1 f_{xw}(X_n(t), W_n(t)) + \alpha_{22}^1 f_{ww}(X_n(t), W_n(t))]$$

$$+ \frac{\lambda_2 Z_2^{(n)}(nt)}{2n}\,[\alpha_{11}^2 f_{xx}(X_n(t), W_n(t)) + 2\alpha_{12}^2 f_{xw}(X_n(t), W_n(t)) + \alpha_{22}^2 f_{ww}(X_n(t), W_n(t))]$$

$$+ O(\tfrac{1}{n} X_n(t)) \quad .$$

The fact that the error term is $O(\frac{1}{n} X_n(t))$ follows from the inequality $Z_i^{(n)}(nt) \le X_n(t)/\nu_i$. From this we also see that (3.16) is $O(X_n(t))$. Since $X_n(t)$ and $W_n(t)$ are martingales it is easy to show that $E(\sup_{t\le T} X_n^2(t))$ and $E(\sup_{t\le T} W_n^2(t))$ are bounded uniformly in n. Consequently Condition (A) of Lemma (1.22) is satisfied. Letting $f_n(t) = f(X_n(t), W_n(t))$ for $f \in C_0^\infty$, $h_n(t)$ is given by (3.16) and since $|h_n(t)| \le CX_n(t)$, (1.27) holds for $p = 2$. Since C_0^∞ is dense in $\hat{C}$, Lemma (1.23) gives tightness for $\{(X_n(t), W_n(t))\}$. Turning to the conditions of Theorem (1.12) we observe that (3.16) can be rewritten as

$$(3.17)\qquad \frac{X_n(t)}{2}\,[a_{11} f_{xx}(X_n(t), W_n(t)) + 2a_{12} f_{xw}(X_n(t), W_n(t)) + + a_{22} f_{ww}(X_n(t), W_n(t))]$$

$$+ \frac{Y_n(t)}{2}\,[b_{11} f_{xx}(X_n(t), W_n(t)) + 2b_{12} f_{xw}(X_n(t), W_n(t)) + b_{22} f_{ww}(X_n(t), W_n(t))]$$

$$+ O(\tfrac{1}{n} X_n(t)) \; ,$$

where

$$a_{ij} = (\lambda_1\mu_2\alpha^1_{ij} - \lambda_2\mu_1\alpha^2_{ij})/(\nu_1\mu_2 - \mu_1\nu_2)$$

(3.18)

$$b_{ij} = (\lambda_2\nu_1\alpha^2_{ij} - \lambda_1\nu_2\alpha^1_{ij})/(\nu_1\mu_2 - \mu_1\nu_2) \ .$$

Let $g(X_n(t), W_n(t))$ denote the coefficient of $Y_n(t)$ in (3.17) and define

(3.19) $$f_n(t) = f(X_n(t), W_n(t)) + \frac{1}{n\eta} Y_n(t)\, g(X_n(t), W_n(t)) \ .$$

Computing $h_n(t)$ we see that

(3.20) $$h_n(t) = Af(X_n(t), W_n(t)) + O(\frac{1}{n} X_n(t), \frac{1}{n} X_n^2(t))$$

and since $E(\sup_{t \leq T} X_n^2(t)) < \infty$, Conditions (1.17) through (1.21) are easily verified. Since C_0^∞ is a core we have weak convergence.

References

[1] Athreya, Krishna B. and Peter E. Ney (1972). Branching Processes. Springer-Verlag, Berlin, Heidelberg, New York.

[2] Billingsley, Patrick (1968). Convergence of Probability Measures. Wiley, New York.

[3] Billingsley, Patrick (1974). Conditional distributions and tightness. Ann. Probability 2, 480-485.

[4] Dynkin, E. B. (1965). Markov Processes. Springer-Verlag, Berlin, Heidelberg, New York.

[5] Feller, William (1951). Diffusion processes in genetics. Proc. Second Berkeley Symp. Math. Statistics Prob. University of California Press, Berkeley, 227-246.

[6] Grimvall, Anders (1974). On the convergence of sequences of branching processes. Ann. Probability 2, 1027-1045.

[7] Helland, Inge S. (1977a). Diffusion approximation of branching processes in random environments. (preprint).

[8] Helland, Inge S. (1977b). Continuity of a class of random time transformations. (preprint).

[9] Jagers, Peter (1971). Diffusion approximations of branching processes. Ann. Math. Statist. 42, 2074-2078.

[10] Jirina, M. (1969). On Feller's branching diffusion processes. Casopis Pest Mat. 94, 84-90.

[11] Keiding, Niels (1975). Extinction and exponential growth in random environments. Theor. Population Biology 8, 49-63.

[12] Kurtz, Thomas G. (1975). Semigroups of conditioned shifts and approximation of Markov processes. Ann. Probability 3, 618-642.

[13] Lamperti, John (1967). The limit of a sequence of branching processes. Z. Wahrscheinlichkeitstheorie und Verw. Gebiete 7, 271-288.

[14] Lindvall, Torgny (1972). Convergence of critical Galton Watson branching processes. J. Appl. Prob. 9, 445-450.

[15] Lindvall, Torgny (1973). Weak convergence of probability measures and random functions in the function space $D[0,\infty)$. J. Appl. Prob. 10, 109-121.

[16] Tanny, David (1977). Limit theorems for branching processes in a random environment. Ann. Probability 5, 100-116.

A MARKOV PROCESS APPROACH TO SYSTEMS OF RENEWAL EQUATIONS WITH APPLICATION TO BRANCHING PROCESSES

by

K.B. Athreya

Indian Institute of Science, Bangalore

and

P. Ney

University of Wisconsin, Madison

§1. INTRODUCTION

This is an expository paper on the asymptotic behavior of systems of renewal equations. Such equations arise in a variety of applications such as semi Markov processes and branching processes. In the latter, the mean and covariance functions typically satisfy such equations. The limiting behavior of such processes is often determined by the asymptotic behavior of the mean and covariance functions.

The simplest case, of course, is the one dimensional renewal equation,

$$m(t) = \zeta(t) + \int_{(0,t]} m(t-u)dF(u), \quad t \geq 0, \tag{1.1}$$

where $F(\circ)$ is a probability distribution function on $(0,\infty)$ and $\zeta(\cdot)$ is a measurable function bounded on finite intervals. The problem is to determine the existence, uniqueness and the asymptotic behavior of the solution $m(\circ)$ to (1.1). The best known theorem for this is available in Feller [9] and says that if $\zeta(\circ)$ is directly Riemann integrable, $F(\circ)$ is non lattice and has a finite mean λ, then $m(\cdot)$ exists, is unique and satisfies

$$\lim_{t\to\infty} m(t) = \lambda^{-1} \int_0^\infty \zeta(u)\,du \tag{1.2}$$

This paper is concerned with proving an analog of (1.2) for a general system of renewal equations (of which (1.1) is a very special case).

The simplest generalization of (1.1) is the "matrix case". Here we are given a $p \times p$ matrix $\{F_{ij}(\cdot)\}_{p\times p}$ of non-decreasing right continuous real valued functions ($F_{ij}(\infty)$ could be infinity), and a p-vector $\{\zeta_i(\cdot)\}$ of measurable functions bounded on finite intervals, all defined on $(0,\infty)$. We then define the system of renewal equations,

$$m_i(t) = \zeta_i(t) + \sum_{j=1}^{p} \int_{(0,t]} m_j(t-u)dF_{ij}(u) \tag{1.3}$$

for $t \geq 0$, $i = 0,1,2,\ldots p$.

The next level is the 'countable case' in which p becomes infinite. Finally the "general case" is defined as follows.

Let $(S, \mathcal{S})$ be a measurable space. Let $\{\mu(s, \cdot)\ s \in S\}$ be a family of measures on $(S \times [0,\infty),\ \mathcal{S} \times \mathcal{B}[0,\infty))$. Let $\zeta(s,t)$ be a measurable function on $S \times [0,\infty)$. Consider the equation

$$(1.4) \quad m(s,t) = \zeta(s,t) + \int_S \int_{(0,t]} m(s', t-u)\, \mu(s, d(s' \times u)).$$

The problem is to determine conditions on $\mu(\cdot,\cdot)$ and $\zeta(\cdot,\cdot)$ to ensure existence and uniqueness of solutions of (1.4), and more importantly to prove an analog of (1.2).

Equation (1.1) arises in a natural way in the theory of Bellman-Harris processes (see Athreya-Ney [4] chapter IV). Equation (1.3) is satisfied by the mean vector in multitype Bellman-Harris processes, multitype Crump-Mode-Jagers processes (see Mode [13], Athreya-Ramamurthy [5]) and in semi-Markov processes with a finite state space (see Cinlar [7], Pyke & Schaufele [17]). Equation (1.4) arises in age dependent branching processes with arbitrary types (Saunders [19]; branching Markov processes (Ikeda et al [10]); branching diffusions (Asmussen & Herring [1]); semi-Markov processes on general state spaces (Cinlar [6], Orey [16], Jacod [11], Athreya, McDonald, Ney [2]).

The analog of (1.2) for (1.3) has been established by Crump [8], Athreya & Ramamurthy [5], Ryan [18], Sevastyanov and Chistyakov [20]. The first two references employ extensions of Feller's compactness arguements and the Choquet-Deny lemma. Ryan uses a clever decomposition of a certain matrix of convolutions to reduce it to the one dimensional case, while the last reference uses Fourier analytic methods under hypotheses which are not minimal.

The general system (1.4) has been treated by Cinlar [6], Crump and Mode (see [13] for references) and Saunders [19]. Cinlar establishes existence and uniqueness in the semi Markov case, i.e., under the restriction that $\mu(s, \cdot)$ is a probability whenever the limit exists. Necessary and sufficient conditions for an operator A to be the infinitesimal generator of a strongly continuous semigroup are given by the Hille-Yosida Theorem (see Dynkin [4] page 30). A subset D of the domain of A, $\mathcal{D}(A)$ is called a core for A if the closure of $A|_D$ (A restricted to D) is A.

<u>Lemma</u> (1.7) Let $T(t)$ be a strongly continuous contraction semigroup on a Banach space K. Then $D \subset \mathcal{D}(A)$ is a core for the infinitesimal operator A if and only if D and $\mathcal{R}(\lambda - A|_D)$ are dense in K for some $\lambda > 0$.

If $D \subset \mathcal{D}(A)$ is dense in K then a sufficient condition for D to be a core is that $T(t) : D \to D$.

We will be interested in the semigroup of a Markov process restricted to a subspace on which it is strongly continuous.

The following lemma gives two examples of cores which we will need in our applications.

<u>Lemma</u> (1.8) (a) $C_0^\infty[0,\infty)$, the space of infinitely differentiable functions with compact support in $[0,\infty)$, is a core for the generator $Af(x) = \alpha x f''(x)$ in $\hat{C}([0,\infty))$, the space of continuous functions vanishing at ∞.
(b) $C_0^\infty([0,\infty) \times (-\infty,\infty))$ is a core for the generator $Af(x,y) = x(af_{xx} + 2bf_{xy} + cf_{yy})$ in $\hat{C}([0,\infty) \times (-\infty,\infty))$.

<u>Proof</u>: We prove part (b). The proof of (a) is similar but much simpler. We begin by observing that the differential equation $u_t = Au$ has solutions of the form $u(t,x,y) = \exp\{-\alpha(t)x + i\beta y\}$ where the real part of $\alpha(t)$ is positive and β is real. In particular

$$(1.9) \qquad \dot{\alpha}(t) = -a\alpha^2(t) + ib\alpha(t)\beta + c\beta^2 ,$$

and we may select $\alpha(0)$ to be any complex number with positive measure for each s. He also shows that if a limit exists then it must necessarily be of a certain type. The last two references use Fourier theoretic arguments employing the Haar Tauberian theorem and the method of Fredholm determinants. A recent paper of Kesten [12] extends Feller type arguments for the case considered by Cinlar [6].

The main objectives of this paper are two fold :

a) To study the system (1.4) in the semi-Markov case using some recent work of Athreya & Ney [3] and Athreya, McDonald and Ney [2] to reduce it to the one dimensional result (1.2) of Feller.

b) To reduce the general system (1.4) to the semi Markov case by an appropriate Malthusian transformation.

Our approach yields the desired limit theory easily and under better hypotheses. It also shows clearly that the asymptotic form of $m(s,t)$ is $e^{\alpha t} v(s) T(\zeta)$ where $v(\circ)$ is a nonnegative function and $T(\circ)$ is a nonnegative linear functional on the space of bounded measurable functions. An unsatisfactory aspect of our approach, as well as of some of the earlier ones, is that the hypotheses are not always readily verifiable in terms of the basic data, namely the family of measures $\{\mu(s,\cdot);\ s \in S\}$. We shall return to this point later. The possibility of the reduction mentioned in (b) is certainly known in the one dimensional case. Although it is doubtless known to those interested in the subject, we did not find explicit mention of this device in the literature either in the 'matrix case' or in the more general case (1.4).

We treat the countable case in section 2, the general case in section 3. The final section discusses applications to branching processes, in particular branching Markov processes, (T_t^0, k, π) processes in the sense of Ikeda, Nagasaw, Watanabe [10], and branching diffusions.

§2. SYSTEMS OF RENEWAL EQUATIONS: THE COUNTABLE CASE.

Let $F_{ij}(\cdot)$ be a matrix (not necessarily finite) of nondecreasing right continuous nonnegative valued functions on $(0,\infty)$, and $\{\zeta_i(\cdot)\}$ be a vector of measurable functions on $[0,\infty)$ that are bounded on finite intervals. In this section we shall study the system

$$(2.1) \qquad m_i(t) = \zeta_i(t) + \sum_j \int_{(0,t]} m_j(t-u) F_{ij}(du)$$

$$i = 1, 2, \dots, \qquad t \geq 0.$$

We shall call the system (2.1) semi-Markov if $P \equiv ((p_{ij}))$, $p_{ij} = F_{ij}(\infty) \equiv \lim_{t\to\infty} F_{ij}(t)$, is the transition probability matrix of a Markov chain.

A semi-Markov system is easily studied via an associated semi-Markov process. Let $\{X_n\}$ be a Markov Chain with P as its transition probability matrix. Conditioned on a realisation $\{X_n = x_n\}_0^\infty$, generate nonnegative random variables $\{L_n\}$ such that

i) the L_n's are independent, and

ii) $P(L_n \leq \ell \mid \{X_n = x_n\}_0^\infty) = \dfrac{F_{x_i, x_{i+1}}(\ell)}{F_{x_i, x_{i+1}}(\infty)}$.

Now construct a continuous time Markov process $\{W(t) = (X(t), A(t)); t \geq 0\}$ by setting

$$(2.2) \qquad W(t) = \begin{cases} (X_0, t) & \text{if } 0 \leq t < L_0 \\ (X_1, t-L_0) & \text{if } L_0 \leq t < L_0 + L_1 \\ (X_2, t-L_0-L_1) & \text{if } L_0+L_1 \leq t < L_0+L_1+L_2 \\ \cdots & \\ (X_n, t-L_0 \cdots - L_{n-1}) & \text{if} \\ \text{and so on.} & L_0+\dots+L_{n-1} \leq t < L_0+L_1+\dots+L_n \end{cases}$$

We can think of $W(t)$ as the state and age since last transition (at time t) of a particle moving as follows. If the particle is in state i, then at the next transition it will jump (in a Markovian manner) to a state j according to the transition probability matrix P. Having determined its next state j, it waits in i for a random length of time with distribution $F_{ij}(\cdot)/F_{ij}(\infty)$ before making the jump to j. The marginal process $\{X(t); t \geq 0\}$ is often referred to as a semi-Markov process since it is Markovian only at the random times $\sum_{0}^{n} L_i$, $n = 0, 1, 2 \ldots$.

We shall now see that certain functionals defined on $\{W(t)\}$ satisfy (2.1) for appropriate $\{\zeta_i(\cdot)\}$, and use this fact together with the ergodicity of $\{W(\cdot)\}$ to study the limiting behavior of (2.1).

Let $f(\cdot, \circ)$ be a bounded measurable function on $\{1, 2, \ldots\} \times [0, \infty)$. Set

$$\overline{m}_i(t) \equiv E_i f(W(t)), \tag{2.3}$$

where E_i stands for expectation under the condition $W(0) = (i, 0)$. (Similarly, P_i is the measure under the condition $W(0) = (i, 0)$.) Then, by decomposing the expectation over the sets $\{L_0 > t\}$ and $\{L_0 \leq t\}$ and using the strong Markov property (a recurrence hypothesis on P is sufficient to ensure that $\{W(t)\}$ is a well defined strong Markov process having no explosions in finite time), we see that

$$\overline{m}_i(t) = f(i, t) P_i(L_0 > t) + \sum_j \int_{(0,t]} \overline{m}_j(t-u) F_{ij}(du) . \tag{2.4}$$

A comparison of (2.1) with (2.4) shows that a solution of (2.1) is given by $\overline{m}_i(t)$ when $f(\cdot, \circ)$ is defined by

$$f(i, t) = \zeta_i(t) \cdot (P_i(L_0 > t))^{-1} . \tag{2.5}$$

If we assume (as we will) that P is irreducible and recurrent then the system (2.1) has a unique solution for all $\zeta_i(\cdot)$ bounded on finite t-intervals. If we further restrict $\zeta_i(\cdot)$ so that $f(\cdot,\cdot)$ as defined by (2.5) is bounded in both arguements, then we could assert that the solution of (2.1) is $\overline{m}_i(t)$. It must be clear now that the asymptotic behavior of solutions to (2.1) is easily studied via the behavior of $E_i f(W(t))$ as $t \to \infty$, i.e. the ergodic behavior of the Markov process $\{W(t);\ t \geq 0\}$. This, in turn, may be studied using the one dimensional result (1.2). In fact, fix a state i_0. Let

$$(2.6)\qquad \begin{cases} N = \inf\{n:\ n \geq 1,\ X_n = i_0\}, \\ T = \sum_{i=0}^{N-1} L_i. \end{cases}$$

Clearly, N is a stopping time for $\{X_n\}$, and T for $\{W(t);\ t \geq 0\}$. For $f(\cdot)$ bounded measurable and $\overline{m}_i(\cdot)$ as defined in (2.4) we get, using the strong Markov property of $\{W(t);\ t \geq 0\}$, the renewal equation

$$(2.7)\qquad \overline{m}_{i_0}(t) = E_{i_0}(f(W(t));\ T > t) + \int_0^t \overline{m}_{i_0}(t-u)\, d_u P_{i_0}(T \leq u).$$

The second term on the right is obtained by using the facts that $W(T) = (i_0, 0)$, and that T and $W(T)$ are independent. (This last point will turn out to be a real difficulty in the general case treated in the next section).

Now (2.7) is a special case of (1.1) with the identification

$$(2.8)\qquad \begin{cases} \zeta(t) = E_{i_0}(f(W(t));\ T > t), \\ F(t) = P_{i_0}(T \leq t). \end{cases}$$

If we ensure that $\zeta(\cdot)$ is d.r.i and $F(\cdot)$ is nonlattice with a finite mean, then from (1.2) it would follow that

(2.9) $$m_{i_0}(t) \longrightarrow \lambda^{-1} \int_0^\infty E_{i_0}(f(W(t)); T > t)\, dt,$$

where $\lambda = E_{i_0}(T)$.

For any other initial distribution of $W(0)$ there is a stopping time τ such that $W(\tau) = (i_0, 0)$, and hence $m_i(t)$ has the same limit for all i. The right side of (2.9) can be identified as follows:

$$\int_0^\infty E_{i_0}(f(W(t)); T > t)\, dt = E_{i_0}\left(\int_0^T f(W(t))\, dt\right)$$

$$= E_{i_0}\left(\sum_{i=0}^{N-1} \int_0^{L_i} f(X_i, u)\, du\right)$$

$$= E_{i_0}\left(\sum_{i=0}^{N-1} \lambda(X_i)\right),$$

where $$\lambda(x) \equiv E_x\left(\int_0^{L_0} f(x, u)\, du\right)$$

$$= \int_0^\infty f(x, u)\, P_x(L_0 > u)\, du.$$

Let $\pi_j \equiv E_{i_0}\left(\sum_{i=0}^{N-1} \delta_{X_i, j}\right)$ be the expected number of visits to j by $\{X_n\}$ between consecutive visits to i_0. Then the limit in (2.9) can be written as

$$(2.10) \qquad \frac{\sum_j \pi_j \int_0^\infty f(j,u)\, P_j(L_0 > u)\, du}{\sum_j \pi_j \int_0^\infty P_j(L_0 > u)\, du} .$$

It is well known that $\{\pi_j\}$ as defined above is an invariant distribution for $\{p_{ij}\}$ and is unique up to a multiplicative constant. Hence, we can use any invariant distribution for $\{p_{ij}\}$ in (2.10) in place of $\{\pi_j\}$.

We must now determine conditions which ensure the direct Riemann integrability of $\zeta(\circ)$ as defined in (2.8). To this end we observe that if for each i, $f(i,t)$ is continuous almost everywhere in t, then $\zeta(t)$ is continuous (a.e.). In fact, let

$$C_1 = \{ t : P_{i_0}(\sum_{i=0}^{n} L_i = t) > 0 \quad \text{for some } n \},$$

$$C_2 = \{ t : f(i,\cdot) \text{ is not continuous at } t \text{ for some } i \}.$$

Clearly C_1 is countable and C_2 has measure zero by assumption. For $t_0 \notin C_1 \cup C_2$, $f(W(t))\, \chi_{(T>t)}$ is continuous at $t = t_0$ w.p.1. under P_{i_0}. By the bounded convergence theorem, $\zeta(t)$ is continuous at t_0. Next, since $\zeta(t)$ is bounded by $P_{i_0}(T>t)$ the d.r.i. of $\zeta(\cdot)$ is implied by $E_{i_0}(T) < \infty$. Summarising the above discussion we arrive at the following

Theorem 2.1. Assume i) The system (2.1) is such that the matrix $P \equiv ((p_{ij} = F_{ij}(\infty)))$ is the transition probability matrix of an irreducible recurrent Markov Chain on $\{1, 2, 3, \ldots\}$. ii) $P_{i_0}(T \le u)$ is nonlattice in u for at least one i_0, where $T = \sum_{i=0}^{N-1} L_i$, with $N = \inf\{ n : n \ge 1, X_n = i_0 \}$.
iii) For some nontrivial invariant distribution $\{\pi_j\}$ for $\{p_{ij}\}$

$$\lambda \equiv \sum_j \pi_j \int_0^\infty P_j(L_0 > u)\, du$$

$$= \sum_j \pi_j \left(\sum_k \int_0^\infty (F_{jk}(\infty) - F_{jk}(u))\, du \right) < \infty .$$

iv) $\zeta_i(\cdot)$ _is continuous_ a.e. _for each_ i .

v) $\sup_{i,t} \zeta_i(t)\,(1 - \sum_j F_{ij}(t))^{-1} < \infty .$

Then, the solution $\{m_i(\cdot)\}$ _to_ (2.1) _exists, is unique, and satisfies_

(2.11) $$\lim_t m_i(t) = \lambda^{-1} \sum_j \pi_j \int_0^\infty \zeta_j(u)\, du .$$

It is possible to give a bootstrap argument to extend the validity of (2.11) without requiring assumption (v) of Theorem 2.1. In the matrix case

$$\zeta_i(t) = \begin{cases} 1 & \text{if} \quad 0 \le t \le h \\ 0 & \text{if} \quad t > 0 \end{cases}$$

for sufficiently small $h > 0$ would satisfy (v), and the solution in this case would give us a Blackwell type renewal theorem, namely $U_{ij}(t+h) - U_{ij}(t) \to h/\lambda$ as $t \to \infty$. Here $U(\cdot)$ is the 'renewal function' defined by

(2.12) $$U(t) = \sum_0^\infty F^{(n)}(t) , \qquad \text{with}$$

$$F^{(n)}(t) = ((F_{ij}^{(n)}(t))) ,$$

$$F_{ij}^{(n)}(t) = \sum_k \int_{(0,t]} F_{ik}^{(n-1)}(t-u)\, dF_{kj}(u) , \quad n \ge 1 ,$$

$$F_{ij}^{(0)}(t) = \begin{cases} \delta_{ij} & \text{if} \quad t > 0 , \\ 0 & \text{if} \quad t \le 0 . \end{cases}$$

From this one can build up (as in Feller [9]) the validity of the above theorem for all directly Riemann integrable $\zeta_i(\cdot)$. When the system is infinite, the argument is more complicated. It does turn out that the above theorem is valid for all $\{\zeta_i(\cdot)\}$ such that $\zeta_i(\cdot)$ is continuous a.e. in t for each i and $\sum_j \pi_j \sum_n \sup_{nh \le t < (n+1)h} |\zeta_j(t)| < \infty$ for sufficiently small $h > 0$. (See Athreya, McDonald, Ney [2] and Cinlar [7]).

We now proceed to reduce a general countable system like (2.1) to a <u>semi Markov</u> one. Systems admitting such a reduction will be called <u>Malthusian</u>.

Assume that the basic data $\{F_{ij}(\cdot)\}$ of (2.1) satisfy:

(2.13) (i) There is a real α such that for all i, j

$$F_{ij}^{*}(\alpha) = \int_0^{\infty} e^{-\alpha u} F_{ij}(du) < \infty$$

and such that

(ii) there is a vector $\{v_i\}$ satisfying

$$v_i > 0 \quad \text{and} \quad \sum_j F_{ij}^{*}(\alpha)\, v_j = v_i \quad \text{for all } i .$$

Now set

$$p_{ij} = \frac{F_{ij}^{*}(\alpha)\, v_j}{v_i} , \tag{2.14}$$

$$\tilde{m}_i(t) = \frac{e^{-\alpha t}\, m_i(t)}{v_i} ,$$

$$\tilde{\zeta}_i(t) = e^{-\alpha t} \frac{\zeta_i(t)}{v_i} ,$$

and

$$G_{ij}(t) = \left(\int_0^{t} e^{-\alpha u} F_{ij}(du) \right) / F_{ij}^{*}(\alpha).$$

Multiplying both sides of (2.1) by $e^{-\alpha t}/v_i$ yields

$$\tilde{m}_i(t) = \tilde{\zeta}_i(t) + \sum_j p_{ij} \int_{(0,t]} \tilde{m}_j(t-u) G_{ij}(du) . \tag{2.15}$$

Clearly, the system (2.15) is semi Markov.

Let $\{X_n\}$ be a Markov chain with transition function $((p_{ij}))$, and as before let $N = N(i_0) = \inf\{n : n \geq 1, X_n = i_0\}$, $T = \sum_{i=0}^{N-1} L_i$, and L_i be a sequence of conditionally (given $\{X_n\} = \{x_n\}$) independent random variables with

$$P(L_i \leq \ell \mid \{X_n\}) = G_{x_i, x_{i-1}}(\ell) .$$

From Theorem 2.1 we can read off the following

<u>Theorem 2.2</u>. <u>Let</u> $\{F_{ij}(\cdot)\}$ <u>of</u> (2.1) <u>satisfy</u> (2.13). <u>Assume</u>

i) $((p_{ij}))$ <u>defined by</u> (2.14) <u>is the transition probability matrix of an irreducible recurrent Markov Chain</u> $\{X_n\}$ <u>on</u> $\{1, 2, 3 \ldots\}$.

ii) $P_{i_0}(T \leq u)$ <u>is nonlattice in</u> u <u>for at least one</u> i_0.

iii) $\lambda \equiv \sum_{j,k} u_j v_j \int_0^\infty \left(\int_t^\infty e^{-\alpha u} F_{jk}(du) \right) dt < \infty$

<u>where the</u> u_j's <u>satisfy</u> $\sum_i u_i F_{ij}^*(\alpha) = u_j$ <u>for all</u> j.

(Such a vector $\{u_j\}$ exists since $\{p_{ij}\}$ is irreducible and recurrent.)

iv) $\zeta_j(\cdot)$ are continuous a.e. for all j, and

$$\sum_j u_j v_j \sum \sup_{nh \leq t < (n+1)h} |\zeta_j(t)| < \infty \quad \text{for all small } h > 0.$$

<u>Then, the solution</u> $\{m_i(\cdot)\}$ <u>of</u> (2.1) <u>exists, is unique, and satisfies</u>

$$e^{-\alpha t} \frac{m_i(t)}{v_i} \longrightarrow \lambda^{-1} \sum_j u_j v_j \int_0^\infty e^{-\alpha u} \zeta_j(u) du .$$

§ 3. SYSTEMS OF RENEWAL EQUATIONS.

Let $(S, \mathcal{S})$ be a measurable space and $\{\mu(s, \cdot); s \in \mathcal{S}\}$ be a family of measures on $\{S \times [0,\infty), \mathcal{S} \times \mathcal{B}[0,\infty)\}$, $\mathcal{B}$ being the Borel sets. Consider the system of renewal equations

$$(3.1) \qquad m(s,t) = \zeta(s,t) + \int_S \int_{(0,t]} m(s', t-u)\, \mu(s, d(s' \times u)),$$

where $\zeta(\cdot,\cdot)$ is a given forcing function. When S is countable, and $\mathcal{S}$ is the subsets of S, then obviously (3.1) reduces to (2.1). We shall call the system (3.1) <u>semi-Markov</u> if $\mu(s, A \times [0,\infty)) \equiv P(s, A)$ is a transition probability function. In this case let $\{X_n ; n \geq 0\}$ be the Markov chain associated with $P(\cdot,\cdot)$.

We assume $\mathcal{S}$ to be countably generated. We can then find a function $G(s, s', t)$ which is jointly measurable in (s, s', t), is a probability distribution in t for fixed (s, s'), and in terms of which (3.1) can be rewritten as

$$(3.2) \qquad m(s,t) = \zeta(s,t) + \int_S \int_{[0,t)} m(s', t-u)\, G(s, s', du)\, P(s, ds').$$

Indeed, $G(s, s', t)$ is simply the Radon-Nikodym derivative of $\mu(s, A \times (0,t])$ with respect to $\mu(s, A \times (0,\infty))$.

We can now associate a Markov process $\{W(t); t \geq 0\}$ with (3.2) just as in the countable case. Conditioned on a realization $\{X_n = x_n\}_0^\infty$ of the underlying chain, generate independent nonnegative random variables $\{L_n\}$ with

$$(3.3) \qquad P(L_i \leq \ell_i \mid \{X_n = x_n\}_0^\infty) = G(x_i, x_{i+1}, \ell_i).$$

Now construct the process $\{W(t) = (X(t), A(t)); t \geq 0\}$ as in (2.2). If the chain $\{X_n\}$ has a recurrence point Δ, i.e.

if $P_x(X_n = \Delta$ for some $n \geq 1) \equiv 1$ for all x, then the method of the previous section carries over to provide a limit theorem for $W(t)$ as $t \to \infty$. In the absence of such a recurrence point the theory gets more complicated. For a discussion of these and other aspects of semi Markov processes on general state spaces see Cinlar [6], Kesten [12], Orey [16], Jacod [11], Nummelin [14] and Athreya, McDonald, Ney [2].

To formulate an analog of theorem 2.1 for the general state space case we need a notion of recurrence for $\{X_n\}$. To this end we will call $\{X_n\}$ recurrent if there exists a set $A \in \mathcal{S}$, a number $\lambda > 0$, and a reference probability measure φ on A, such that

$$(3.4) \quad \begin{cases} \text{(i)} & P_x(X_n \in A \text{ for some } n \geq 1) = 1 \\ \text{(ii)} & P(x, E) \geq \lambda \varphi(E) \text{ for all } x \in A, \ E \subset A. \end{cases}$$

This definition is closely related to Harris-recurrence (see [3]). Under this condition it can be shown [3] that there always exists an invariant measure for P, namely a measure $\pi(\cdot)$ such that

$$\pi(\cdot) = \int_S P(x, \cdot)\, \pi(dx).$$

If we further assume that $G(x, x', t) \equiv G(x, t)$ is independent of x' (as we do from now on), then we have the following result (see [2]).

Theorem 3.1. Assume that $\{X_n\}$ is recurrent in the sense of (3.4), that $G(x,t)$ is nonlattice in t for all x, and that

$$\theta = \int_S \int_0^\infty (1 - G(x,t))\, dt\, \pi(dx) < \infty .$$

Then for any initial distribution μ *of* $W(0)$

$$(3.5)\qquad \lim_{t\to\infty} P_{\mu}\{W(t) \in B\times[a,b]\} = \theta^{-1}\int_B (dx)\int_a^b (1-G(x,t)dt.$$

If furthermore the forcing function ζ *satisfies*

$$\pi\,(x : \zeta\,(x,t) \text{ is discontinuous for some } t\} = 0$$

and

$$\int \pi\,(dx) \sum_{n=0}^{\infty} \sup\{\,|\,\zeta\,(x,t)| : nh \le t < (n+1)h\} < \infty$$

for some $h > 0$, *then for all* $s \in S$ *the solution of* (3.1) *satisfies*

$$(3.6)\qquad \lim_{t\to\infty} m(s,t) = \theta^{-1}\int_S \pi\,(dx)\int_0^{\infty} \zeta(x,t)\,dt.$$

Finally we carry out the reduction of a general system (3.1) to a semi-Markov system. We assume that there exists a real α, and a function $v(\cdot)$ on $(S,\mathcal{S})$ to $(0,\infty)$ such that

$$(3.7)\qquad \int_S v(s')\,\mu_{\alpha}(s,ds') = v(s),$$

where

$$\mu_{\alpha}(s,A) = \int_0^{\infty} e^{-\alpha u}\, d_u\,\mu(s, A\times[0,u]).$$

Set

$$(3.8)\quad \begin{cases} \tilde{m}(s,t) = e^{-\alpha t}\, m(s,t)/v(s)\,, \\ \tilde{\zeta}(s,t) = e^{-\alpha t}\, \zeta(s,t)/v(s)\,, \\ P(s,t) = \int_S \frac{v(s')}{v(s)}\, \mu_\alpha(s,ds')\,, \\ \mu_\alpha(s,t,A) = \int_A \int_{(0,t]} e^{-\alpha u}\, \mu(s, d(s' \times u))\, \frac{v(s')}{v(s)}\,. \end{cases}$$

Since $\mu_\alpha(s,t,A)$ increases to $\mu_\alpha(s,A)$ as $t \to \infty$, there exist Radon-Nikodym derivatives $G_\alpha(s,s',t)$ jointly measurable in (s,s',t) such that (3.1) becomes

$$(3.9)\quad \tilde{m}(s,t) = \tilde{\zeta}(s,t) + \int_S \int_{[0,t]} \tilde{m}(s', t-u)\, G_\alpha(s,s',du)\, P(s,ds').$$

Clearly, (3.9) is a semi-Markov system. Any system (3.1) for which a real α and a positive $v(\circ)$ satisfying (3.7) exist will be called <u>Malthusian</u>, and α, the Malthusian parameter. If the system (3.9) satisfies the hypotheses of Theorem 3.1 then $\tilde{m}(s,t)$ will have a limit independent of s and of the form

$$(3.10)\quad \int_S \pi(dx) \int_0^\infty \tilde{\zeta}(s,u)\, du \equiv T\zeta \qquad \text{(say)}\ .$$

Thus, in the Malthusian case, when $\tilde{m}(\circ,\circ)$ has a limit, $m(s,t)$ is of the form $e^{\alpha t}\, v(s)\, T\zeta$ for large t.

§ 4. APPLICATIONS TO BRANCHING PROCESSES

4.1. General discussion:

As remarked earlier, systems of renewal equations of the type (3.1) are satisfied by the means and higher moments of a large class of branching processes. We shall now derive such an equation for a fairly general branching model that includes, among others, single and multitype Galton-Watson processes, single and multitype age dependent branching processes, multitype Crump-Mode-Jagers processes, Sevastyanov-processes, (T_t, k, π) process of Ikeda et al.

Let $(S, \mathcal{S})$ be a measurable space. Consider a branching process with S as its type space and evolving as follows. All particles move in S during their life time as a Markov process and generate particles of various types at random times before their death. Each new entrant evolves on its own and independently of the others. To be more specific we are given:

i) A Markov process $\{X_t\}$ with a random life time τ and state space S.

ii) A measure valued off-spring process $\{\eta(s,t);\ t \geq 0\}$, $\eta(s,t)(A)$ denoting the number of particles with type in A created during the interval $(0, t]$ by a new particle of type s created at time 0.

To obtain a Markovian description of the above process we need to obtain for each t the state chart $Y_t = \{\theta_1, \theta_2, \ldots, \theta_{Z_t}\}$, where Z_t is the number of particles alive at t, and each $\theta_i = (s_i, \varphi_i)$, s_i being the location and φ_i being the offspring history, including age, of the ith particle. If the offspring process $\{\eta(s,t);\ t \geq 0\}$ has the stationary Markovian property that

$$\{\eta(s, t+u);\ t \geq 0 \mid \tau > u,\ X(u)\} \stackrel{d}{=} \{\eta(X(u), t);\ t \geq 0\},$$

as is the case for a (T_t, k, π) process or a Bellman-Harris process with age as type, then the position chart $\{s_1, s_2, \dots s_{Z_t}\}$ itself is Markovian. Let Θ be the totality of all possible θ's and $\mathcal{B}$ a σ-algebra on Θ induced by S, and rich enough to make the offspring processes measurable. Let $f : \Theta \to R$ be bounded and measurable. Set

$$(4.1) \qquad m(\theta, t) = E_\theta \left\{ \sum_{i=1}^{Z_t} f(\theta_i) \right\},$$

where E_θ stands for the expectation operator with $Y_0 = \theta$.

Since the process $\{Y_t\}$ is stationary Markovian we see from the basic description of the process that $m(\cdot)$ satisfies

$$(4.2) \qquad m(\theta, t) = E_\theta \{ f(X_t); \tau > t \} + \int_S \int_{(0,t]} m(s', t-u)\, \mu(\theta, d(s' \times u)),$$

where $m(s,t)$ stands for $m(\theta, t)$ when θ denotes the state of a brand new particle of type s and $\mu(\theta, A \times (0,u])$ is the mean number created in $(0,u]$ with type in A by a single particle in state θ at time 0. If we restrict θ to new particles then (4.2) is of the form (3.1) with

$$\zeta(s, t) = E\{ f(X_t); \ \tau > t \mid X_0 = s \}.$$

The asymptotic behavior of $m(\theta, t)$ for general θ follows from that of $m(s,t)$ via the equation (4.2).

4.2. Specialization to the (T_t, k, π) model.

This model, introduced by Ikeda et al [10], includes as special cases multitype Bellmann-Harris processes, branching diffusions etc. In this model, each particle moves in a space $(S, \mathcal{S})$ according to a Markov process with semigroup $\{T_t\}$, with a random life time governed by the killing density $k(\circ)$

(i.e. $P\{$ particle dies in $(t, t+h) \mid$ it is alive at t and is at $s\}$ $= k(s)h + o(h)$). At the end of its life time τ it produces a random number of offspring distributed over $(S, \mathcal{S})$ with distribution $\pi(X(\tau), \cdot)$, depending on $X(\tau)$, the location of the parent particle at its death. If all the offspring are concentrated at $X(\tau)$ then the branching is called <u>local</u>. The renewal equation (4.2) now becomes

$$(4.3) \qquad m(s,t) = E\{f(X_t);\ \tau > t \mid X_0 = s\}$$

$$+ E\{ (\int_S m(s', t-\tau)\, \pi(X(\tau), ds'));\ \tau \le t \mid X_0 = s\}.$$

This is of the form (3.1) with

$$(4.4) \quad \begin{cases} \zeta(s,t) = E\{ f(X_t);\ \tau > t \mid X_0 = s\} \\ \mu(s, A \quad (0,t]) = E\{ \int_S \chi_A(s')\, \pi(X(\tau), ds');\ \tau \le t \mid X_0 = s\}. \end{cases}$$

The system is Malthusian if we can find a real α and a positive function $v(\circ)$ on $(S, \mathcal{S})$ such that

$$(4.5) \qquad \int v(s')\, \mu_\alpha(s, ds') = v(s),$$

where

$$\mu_\alpha(s, A) = \int_0^\infty e^{-\alpha u}\, d_u \mu(s, A \times (0,u])$$

$$= E\{e^{-\alpha\tau} \int_S \chi_A(s')\, \pi(X(\tau), ds') \mid X_0 = s\}.$$

Thus, (4.5) becomes

$$(4.6) \qquad E\{ e^{-\alpha\tau} \int_S v(s')\, \pi(X(\tau), ds') \mid X_0 = s\} = v(s).$$

If we set

$$(4.7) \qquad (\pi h)(s) = \int_S h(s')\,\pi(s, ds'),$$

then (4.6) becomes

$$(4.8) \qquad E\{e^{-\alpha\tau}(\pi v)(X(\tau)) \mid X_0 = s\} = v(s).$$

By the definition of a (T_t, k, π) process, the left side of (4.8) becomes

$$E\{\int_0^\infty e^{-\alpha u}(\pi v)(X(u))\,k(X(u))\,e^{-\int_0^u k(X(u))du}\,du \mid X_0 = s\}.$$

To simplify this introduce the Markov process

$$\tilde{X}(t) = (X(t), \int_0^t k(X(u))\,du),$$

with state space $\tilde{S} = S \times [0, \infty)$, generator $\tilde{A}$, and resolvent operator $\tilde{R}_\alpha$. Then if $\alpha \geq 0$ and f is in the domain of $\tilde{R}_\alpha$,

$$(4.9) \qquad \begin{aligned}(\tilde{R}_\alpha f)(s, s') &= \int_0^\infty \tilde{T}_t f(s, s')\,e^{-\alpha t}\,dt \\ &= \int_0^\infty E_{(s, s')}\,f(X(t), s' + \int_0^t k(X(\theta))d\theta)\,e^{-\alpha t}\,dt.\end{aligned}$$

Definning $f: \tilde{S} \to R$ by

$$(4.10) \qquad f(s,t) = (\pi v)(s)\,k(s)\,e^{-t}$$

we see that

$$(\tilde{R}_\alpha f)(s, s') = \int_0^\infty E_{(s, s')}(\pi v)(X(t))\, k(X(t))\, e^{-s' - \int_0^t k(X(\theta))d\theta} e^{-\alpha t} dt,$$

(4.11)

$$= e^{-s'}(\tilde{R}_\alpha f)(s, 0) = e^{-s'} v(s),$$

since by (4.8)

(4.12) $$(\tilde{R}_\alpha f)(s, 0) = E_{(s, 0)} \{ e^{-\alpha\tau} (\pi v)(X(\tau)) \} = v(s) .$$

From the identity

$$(\alpha - \tilde{A})\tilde{R}_\alpha = \tilde{R}_\alpha(\alpha - \tilde{A}) = I$$

we see that

$$(\alpha - \tilde{A})(\tilde{R}_\alpha f)(s, s') = f(s, s'),$$

and hence by (4.10) and (4.11)

(4.13) $$(\alpha - \tilde{A})(e^{-s'} v(s)) = (\pi v)(s)\, k(s) e^{-s'}.$$

But for $g \in \mathcal{D}(\tilde{A})$

$$(\tilde{A} g)(s, s') = k(s) \frac{\partial g}{\partial s'}(s, s') + (Ag)(s, s'),$$

and applying this to (4.13) we are led to the eigenvalue problem

(4.14) $$(Av)(s) + k(s)(\pi v - v)(s) = \alpha v(s)$$

Asmussen and Herring [1] arrived at (4.14) for the case of a branching diffusion by a different route. They then make a direct analysis of the eigenvalue problem (4.14), and use the completeness of the associated eigenfunctions to expand the solution $m(s, t)$ in terms of these eigenfunctions to reach the conclusion that $e^{-\alpha t} m(s, t) \to v(s) T\zeta$. In our approach we have to verify the conditions needed for the ergodicity of the semi Markov process associated with the reduced process. The translation of these conditions in terms of the original data, viz (T_t^0, k, π) still needs to be carried out.

REFERENCES

[1] Asmussen, S. and Herring, H. (1976) Strong limit theorems for general supercritical branching processes with applications to branching diffusions. Z. Wahrschenilichkeitstheorie 36 (195-212).

[2] Athreya, K.B., McDonald, D., and Ney, P. Limit theorems for semi-Markov processes and renewal theory for Markov chains. To appear in Annals of Probability.

[3] Athreya, K.B., and Ney, P. A new approach to the limit theory of recurrent Markov chains. To appear in Trans. A.M.S.

[4] Athreya, K.B., and Ney, P. (1972) Branching Processes. Springer Verlag, Heidelberg.

[5] Athreya, K.B., and Ramamurthy, K. (1977) Convergence of state distribution in supercritical multitype Crump-Mode-Jagers processes. J. of the Indian Math. Soc.

[6] Cinlar, E. (1969) On semi-Markov processes on arbitrary state spaces. Proc. Camb. Phil. Soc., 66 (381-392).

[7] Cinlar, E. (1975) Introduction to Stochastic processes. Prentice Hall, New Jersey.

[8] Crump, K. (1970) On systems of renewal equations. J. of Math. Anal. and Appl's., 30 (425-434).

[9] Feller, W. (1971) An Introduction to Probability Theory and its Appl's. Vol. 2, 2nd ed., Wiley, N.Y.

[10] Ikeda, N., Nagasawa, M., and Watanabe, S. (1968-69) Branching Markov processes I, II, III. J. of Math of Kyoto Univ. 8 (233-278, 365-410), 9(95-160).

[11] Jacod, J. (1971) Théorème de renouvellement et classification pour les chaînes de semi-Markoviennes. Ann. Inst. Henri Poincaré. Section B7 (83-129).

[12] Kesten, H. (1974) Renewal theory for Markov chains. Ann. Prob. 2 (355-387).

[13] Mode, C.J. (1971) Multitype Branching Processes. American Elsevier, New York.

[14] Nummelin E. Uniform and ratio limit theorems for Markov renewal and semi-regenerative processes on a general state space. Helsinki Univ. of Tech. report.

[15] Orey, S. (1971). Limit Theorems for Markov Chain Transition Probabilities. Van Nostrand, New York.

[16] Orey, S. (1961). Change of time scale for Markov processes. Trans. A.M.S. 99 (384-397).

[17] Pyke, R. and Schaufele, R. (1964). Limit theorems for Markov renewal processes. Ann. Math. Stat. 35 (1746-1764).

[18] Ryan, T. (1976) A multidimensional renewal theorem. Ann. Prob. 4 (656-661).

[19] Saunders, D.I. (1976). Malthusian behavior of the critical and subcritical age-dependent branching process with arbitrary state space. J. Appl. Prob. 13 (455-465).

[20] Sevastyanov, B.A. (1971) Renewal type equations and the moments of branching processes (Russ.) Math. Zametki 3 (3-14).

INDEX